Global Oil and Gas Resources: Potential and Distribution

Lirong Dou · Zhixin Wen · Zhaoming Wang

Global Oil and Gas Resources: Potential and Distribution

With Contributions by Lirong Dou, Zhixin Wen, Zhaoming Wang, Zhengjun He, Chengpeng Song, Ruiyin Chen, Kun Liang, Xiaobing Liu, Zuodong Liu, Haiguang Bian, Xueling Wang, Yonghua Wang, Tianyu Ji, Hengxuan Li, Xi Chen, Tiansi Luan, Ling Su, Yu Ji

石油工业出版社有限公司
PETROLEUM INDUSTRY PRESS

Springer

Lirong Dou
Research Institute of Petroleum Exploration
and Development, China National
Petroleum Corporation
Beijing, China

Zhixin Wen
Research Institute of Petroleum Exploration
and Development, China National
Petroleum Corporation
Beijing, China

Zhaoming Wang
Research Institute of Petroleum Exploration
and Development, China National
Petroleum Corporation
Beijing, China

ISBN 978-981-97-4755-9   ISBN 978-981-97-4756-6   (eBook)
https://doi.org/10.1007/978-981-97-4756-6

Jointly published with Petroleum Industry Press
The print edition is not for sale in China (Mainland). Customers from China (Mainland) please order the print book from: Petroleum Industry Press.
ISBN of the Co-Publisher's edition: 9787518348367

This work was supported by Petroleum Industry Press.

This Springer imprint is published by the registered company Springer Nature Singapore Pte Ltd.
The registered company address is: 152 Beach Road, #21-01/04 Gateway East, Singapore 189721, Singapore

If disposing of this product, please recycle the paper.

Foreword

Since the 1950s, China's petroleum industry has undergone rapid development over seven decades, establishing a comprehensive global oil and gas industry chain with an extensive workforce and equipment, making it one of the world's leading oil and gas producers. By 2023, China produced 208 million tons of crude oil (6th globally) and 230 billion cubic meters of natural gas (4th globally). Concurrently, China is a significant consumer of oil and gas, consuming 705 million tons of oil in 2023, with a 77.6% reliance on foreign sources; and 395.3 billion cubic meters of natural gas, with a 42.3% dependency on imports. The natural endowment of oil and gas resources in China, along with the volume of oil and gas consumption, and the future development of the oil and gas sector, underscores the urgent need for China to broaden its presence in the international oil and gas market. Engaging proactively in the development and utilization of overseas oil and gas resources is vital for safeguarding China's energy security and contributes significantly to the prosperity of the global oil and gas sector.

China National Petroleum Corporation (CNPC) is China's largest oil company, playing a pivotal role in the integration of China's oil sector with the global industry. Since initiating its strategy of overseas oil and gas cooperation in 1993, CNPC has achieved remarkable success over 31 years, particularly in oil and gas cooperation under the Belt and Road Initiative. By the end of 2023, CNPC had overseen 83 oil and gas projects across 34 countries in five key oil and gas cooperation regions: the Middle East, Central Asia and Russia, Africa, the Americas and the Asia-Pacific, with its overseas oil and gas equity production surpassing 100 million tons of oil equivalents.

Comprehending the status of global oil and gas resources, effectively assessing the potential for reserve growth of discovered oil and gas fields worldwide, and scientifically predicting the potential and distribution of undiscovered recoverable oil and gas resources are fundamental for conducting international oil and gas cooperation activities. Renowned international oil conglomerates, including ExxonMobil, Shell and BP, have consistently regarded the global distribution of oil and gas resources as a critical strategic foundation. Their systematic research and periodic release of evaluation outcomes serve as vital references for oil companies participating in global oil and gas production and in optimizing their asset portfolios.

Since 2015, the global oil and gas resources evaluation research team of the Research Institute of Petroleum Exploration and Development, CNPC (RIPED), has been actively involved in a comprehensive global oil and gas geological assessment. They have conducted a scientific evaluation and forecast of the total oil and gas resources for the upcoming three decades across 468 basins globally. The evaluation encompasses the remaining recoverable reserves, potential for growth and distribution of 34,119 oil and gas fields, the undiscovered recoverable resource potential and distribution of 920 conventional oil and gas plays, and the recoverable resource potential and distribution of 656 unconventional plays, which include seven types of unconventional resources such as tight gas, shale gas, coalbed methane, tight oil, oil shale, heavy oil, and oil sands.

This publication is a systematic summary of these significant research findings, signifying a new breakthrough in the scientific comprehension of global oil and gas resources by Chinese oil companies. It addresses the limitations of international oil companies and authoritative bodies that typically disclose evaluation outcomes without the ability to verify their precision. The research findings will serve as crucial references for the formulation of China's energy strategies and oil and gas policies. Furthermore, they provide a valuable reference for oil companies across the globe as they participate in global oil and gas production cooperation.

August 2024   Desheng Li
                                                                           Member of the Chinese Academy
                                                                               of Sciences, AAPG Distinguished
                                                                                         Achievement Award

Preface

Oil and gas remain among most crucial primary energy sources for consumption. The development of low-carbon and non-carbon energy sources is a prevailing trend in global energy development, but it will not be achieved quickly. Many domestic and foreign agencies predict that even in the most optimistic scenario, oil and gas consumption will still account for approximately 20% of the world's primary energy consumption in 2050. Equal consideration must be given to promoting low-carbon development and ensuring energy security. Therefore, the quantity and distribution of oil and gas resources worldwide, as well as their efficient utilization in a low-carbon manner, are crucial scientific issues.

In 2008, the project "Technology for Rapid Evaluation of Remaining Global Oil and Gas Resources and Assets" was initiated under a major national oil and gas research program titled "Development of Large Oil and Gas Fields and CBM." Through continuous research efforts spanning over ten years, the RIPED has developed a technology system for evaluating oil and gas resources based on "oil and gas accumulation play" and has successfully completed the assessment of global oil and gas resources with the broadest scope and the most comprehensive range of oil and gas resources.

This book was compiled by a team of experts and technicians from the RIPED Overseas Research Center. Lirong Dou and Zhixin Wen presented the general concept of compilation and conducted a comprehensive review of the entire book. The Preface, Introduction, and Chap. 1 of this book were written by Lirong Dou, Zhixin Wen, and Zhaoming Wang. Chapter 2 was written by Zhaoming Wang and Ruiyin Chen. Chapter 3 was written by Lirong Dou, Zhixin Wen, and Zhaoming Wang. Chapter 4 was written by Haiguang Bian, Ruiyin Chen, and others. Chapter 5 was written by Haiguang Bian, Xi Chen, and others. Chapter 6 was written by Xiaobing Liu, Xueling Wang, and others. Chapter 7 was written by Chengpeng Song and others. Chapter 8 was written by Xiaobing Liu, Hengxuan Li, and others. Chapters 9 and 10 were written by Zhengjun He and others. Chapter 11 was written by Zuodong Liu, Tiansi Luan, Kun Liang and others. Chapter 12 was written by Zhixin Wen, Zhaoming Wang, Lirong Dou, and others. Finally, Lirong Dou, Zhixin Wen, Zhaoming Wang and Yu Ji completed the general compilation, revision, and finalization of the entire text of this book.

Special thanks go to Academician Xiaoguang Tong for overseeing the evaluation and research of global oil and gas resources since the "11th Five-Year Plan" period. He has formulated detailed research ideas and plans, cultivated a global oil and gas resources evaluation team, led, and guided scientific researchers to deepen this research continuously, and provided ongoing technical support for PetroChina to expand its overseas oil and gas business. We would like to express our sincere thanks to Prof. Jiping Zhou and CNPC for their strong support. We also extend our gratitude to Prof. Jintao Chen, Prof. Wenyuan He, Prof. Xiandeng Ye, Prof. Yong Jia, Prof. Liangqing Xue, Prof. Henian Liu, Prof. Xiaohua Pan, and other leaders and experts for their valuable guidance. We are grateful to S&P Global and Wood Mackenzie for their support in providing relevant data and information.

Even though this book has been written to the best of our knowledge and understanding, there will inevitably be imperfections due to level- and data-related limitations. We sincerely hope that readers can offer valuable opinions and suggestions to help us continuously improve our research and work in the future.

Beijing, China   Lirong Dou
March 2024   Zhixin Wen
                                                                                              Zhaoming Wang

Disclaimer

The sources of all data and views contained in this report are deemed to be reliable, but the authors make no warranty as to the accuracy or completeness of such data and/or views. The information contained herein is for reference purposes only. No information contained herein, or views expressed herein shall constitute the basis for making any quotation or request for quotation of any involved securities transaction or for making other investment decisions. The authors are not responsible or liable for any losses incurred due to the use of any information contained herein unless expressly provided by law.

Readers should not rely on this report as a substitute for their independent judgment or make decisions solely based on it. The authors may issue or publish other reports containing information that is inconsistent with the data contained herein and conclusions that differ from those contained herein. This report reflects the diverse views, insights, and analytical methods of researchers and does not represent the official position of the authors.

The information, opinions, and predictions contained herein only reflect the researchers' judgment as of the date of issue of this report and are subject to change from time to time without notice. No organization or individual shall copy, reproduce, publish, reprint, or cite this report in any form without the prior written consent or permission of the authors. Otherwise, all adverse consequences and legal liabilities resulting from the copying, reproduction, publication, reprinting, or citation of this report shall be assumed by the organization or individual engaging in these actions without proper consent or permission.

About This book

This book contains the basic data of a number of basins, countries, and regions with independent intellectual property rights, which has been acquired through a new round of potential reevaluation of the conventional and unconventional oil and gas resources in 468 basins worldwide conducted based on new oil and gas exploration discoveries, new data and new geological insights using appropriate, applicable evaluation methods on the basis of a major national research project of oil and gas science and technology titled "Research on the Evaluation of Global Oil and Gas Resources and Selection of Oil and Gas-bearing Plays."

This book provides reference and inspiration for countries and enterprises to analyze the supply and demand relationship of oil and gas, assess the future trend, formulate relevant policies, and seek opportunities for oil and gas cooperation. It can also be used as a reference for teaching in universities and research institutes.

Introduction

The assessment of the geology and potential of global oil and gas resources is the foundation for international cooperation in the global oil and gas industry. Large international oil companies and research institutions have conducted independent studies, typically treat the results as confidential and do not disclose them to the public. The United States Geological Survey (USGS) assesses oil and gas resources both within the United States and in select basins worldwide, focusing on petroleum systems. The evaluation results are periodically disclosed to the public. The International Energy Agency (IEA) publishes annual energy outlooks, while Beyond Petroleum (*BP*) provides yearly updates on the status of global oil and gas reserves and consumption. The data disclosed by them serve as the foundation for analyzing the international oil and gas exploration potential, as well as oil and gas supply and demand, and for developing strategies related to oil and gas. Therefore, it is necessary to evaluate the potential of global oil and gas resources independently from Chinese perspectives, based on China's national oil and gas development strategy. This evaluation should utilize methods appropriate to Chinese characteristics to obtain results with independent intellectual property rights.

Since 2008, through continuous efforts and by leveraging major national oil and gas research projects and the significant science and technology projects of CNPC, RIPED has developed an innovative method for assessing conventional and unconventional oil and gas resources through the concept of "oil and gas accumulation play." The institute has conducted a comprehensive evaluation of the geology and potential of conventional oil and gas resources and seven types of unconventional resources in the primary oil and gas-rich basins globally. As a result, RIPED has acquired evaluation data with exclusive intellectual property rights for the first time. The evaluation data was disclosed to the public for the first time in 2017, addressing a significant domestic information gap. The disclosure attracted extensive attention from the public, and the disclosed data was widely used. Supported by continuous funding support from the National Oil and Gas Research Project 29 titled "Research on the Evaluation of Global Oil and Gas Resources and Selection of Oil and Gas-bearing Plays" during the "13th Five-Year Plan" period, RIPED established a new mechanism for progressive evaluation, continuous updating and regular disclosures.

For the results and data disclosed herein, based on the evaluation of global oil and gas resources completed during the "13th Five-Year Plan" period, new discoveries, data, and new insights from oil and gas exploration, as well as the formation conditions of conventional and unconventional oil and gas reservoirs, the criteria for classifying conventional and unconventional reservoirs have been re-determined. A set of criteria and processes have been established for screening all types of conventional and unconventional oil and gas resources, considering each basin as a whole and each petroleum system as a core. A system of methods is used to evaluate all types of conventional and unconventional oil and gas resources by "oil and gas accumulation play" and applicable to varying extents of exploration has been developed. Using these criteria, processes, and methods, the RIPED has reevaluated 468 basins worldwide, established the first platform for managing fundamental data on potential traps globally, and completed the reevaluation and reidentification of global conventional and unconventional oil and gas resources' potential and distribution.

There are four parameters for evaluating global conventional oil and gas resources: cumulative production, remaining recoverable reserves, reserve growth from discovered oil and gas fields, and undiscovered recoverable resources. Cumulative production and remaining recoverable reserves in the 468 basins have been recalculated and updated based on the relevant data provided by S&P Global. The potential for reserve growth from discovered oil and gas fields in the next 30 years has been recalculated using the reserve growth function. The undiscovered recoverable oil and gas resources in 468 basins have been reevaluated and updated based on new discoveries, new data, and new insights. For unconventional oil and gas resources, the technically recoverable reserves of shale oil and gas in 260 plays distributed in 132 overseas basins have been reevaluated and updated as a key index. The latest evaluation data recently published by Chinese scholars was mainly referred to the resource potential of Chinese basins. Therefore, this dataset includes both conventional and unconventional data from all basins worldwide, making it the most comprehensive. The complete dataset includes the remaining recoverable reserves of 34,119 oil and gas fields and the potential for reserve growth from these fields, the undiscovered recoverable resources in 920 plays, and the technically recoverable reserves of seven types of unconventional oil and gas in 656 plays, including shale oil/tight oil, oil shale, heavy oil, oil sands, tight gas, shale gas, and Coal Bed Methane (CBM).

According to the latest evaluation results, the global recoverable reserves of conventional oil and gas are $11,434.9 \times 10^8$ tons of oil equivalent, and the global recoverable reserves of conventional oil are 5292.8×10^8 tons of oil equivalent, accounting for 46.3% of the global recoverable reserves of conventional oil and gas; the recoverable reserves of condensates are 581.4×10^8 tons, accounting for 5.1%; and the recoverable reserves of gas are 650.6×10^{12} m^3, accounting for 48.6%. Conventional oil and gas resources are distributed in regions including the Middle East, Russia, Asia-Pacific, Central and South America, Africa, North America, Central Asia, and Europe (these regions are listed in descending order depending on the volumes of oil and gas resources). As of the end of 2023, the cumulative oil and gas production is 2608×10^8 tons of oil equivalent, representing a recovery rate of 22.8%. This indicates that the remaining conventional oil and gas resources still have significant potential for exploration and development (Fig. 1 and Table 1).

Remaining recoverable reserves of oil and gas represent the potential for future development. As of the end of 2023, the global remaining recoverable oil and gas reserves are 3908.1×10^8 tons of oil equivalent, accounting for 34.2% of the global recoverable reserves of conventional oil and gas resources. The remaining recoverable oil and gas reserves are primarily distributed in Russia, Qatar, Saudi Arabia, and other countries. They are concentrated in the Arabian Basin, West Siberian Basin, Zagros Basin, and others. The remaining recoverable reserves of oil and gas in onshore and offshore reservoirs account for 54.4% and 45.6% of the total remaining recoverable reserves, respectively. The remaining recoverable oil and gas reserves are mainly distributed in foreland basins, passive margin basins, and rift basins, spanning the Cretaceous, Jurassic, Permian, Paleogene, and Neogene formations.

Reserve growth from discovered oil and gas fields can represent the potential for progressive exploration in the future. The global reserve growth potential of discovered oil and gas fields are 1119.6×10^8 tons of oil equivalent, accounting for 9.8% of the global recoverable reserves of conventional oil and gas resources. The potential oil and gas resources are mainly distributed in Russia, Saudi Arabia, Qatar, and other countries and distributed in the Arabian Basin, Zagros Basin, West Siberian Basin, and others. The potential oil and gas resources in onshore and offshore reservoirs account for 56.4% and 43.6%, respectively. These oil and gas resources are distributed in foreland basins, passive margin basins, and rift basins, and the Cretaceous, Jurassic, Paleogene, Permian, and Neogene formations.

Undiscovered recoverable oil and gas resources mainly represent the future exploration potential and direction. The global undiscovered recoverable oil and gas resources are 3799.2×10^8 tons of oil equivalent, accounting for 33.2% of the global recoverable conventional oil and gas resources. These undiscovered recoverable resources are mainly distributed in Russia, China, Brazil, the United States, and other countries and concentrated in the Arabian Basin, West Siberian Basin, Zagros Basin, Santos Basin, and others. The undiscovered recoverable resources in onshore and offshore reservoirs account for 59.7% and 40.3%, respectively. These resources are distributed in basins in passive continental margins, foreland basins, rift basins, and the Cretaceous, Jurassic, Permian, Paleogene, and Neogene formations.

The results of this evaluation showed that the technically recoverable unconventional oil and gas resources worldwide are 7249.9×10^8 tons of oil equivalent, of which the technically recoverable unconventional oil resources are 4244.9×10^8 tons, accounting for 58.6%, and the technically recoverable unconventional gas resources are 351.6×10^{12} m^3, accounting for 41.4%. These technically recoverable resources are distributed across various regions, including North America, Russia, Central and South America, Asia-Pacific, Europe, the Middle East, Africa, and Central Asia. With the continuous progress of cost-effective oil and gas development technologies, unconventional oil and gas resources, especially shale oil and gas, will become realistic alternative resources in the future (Fig. 2 and Table 2).

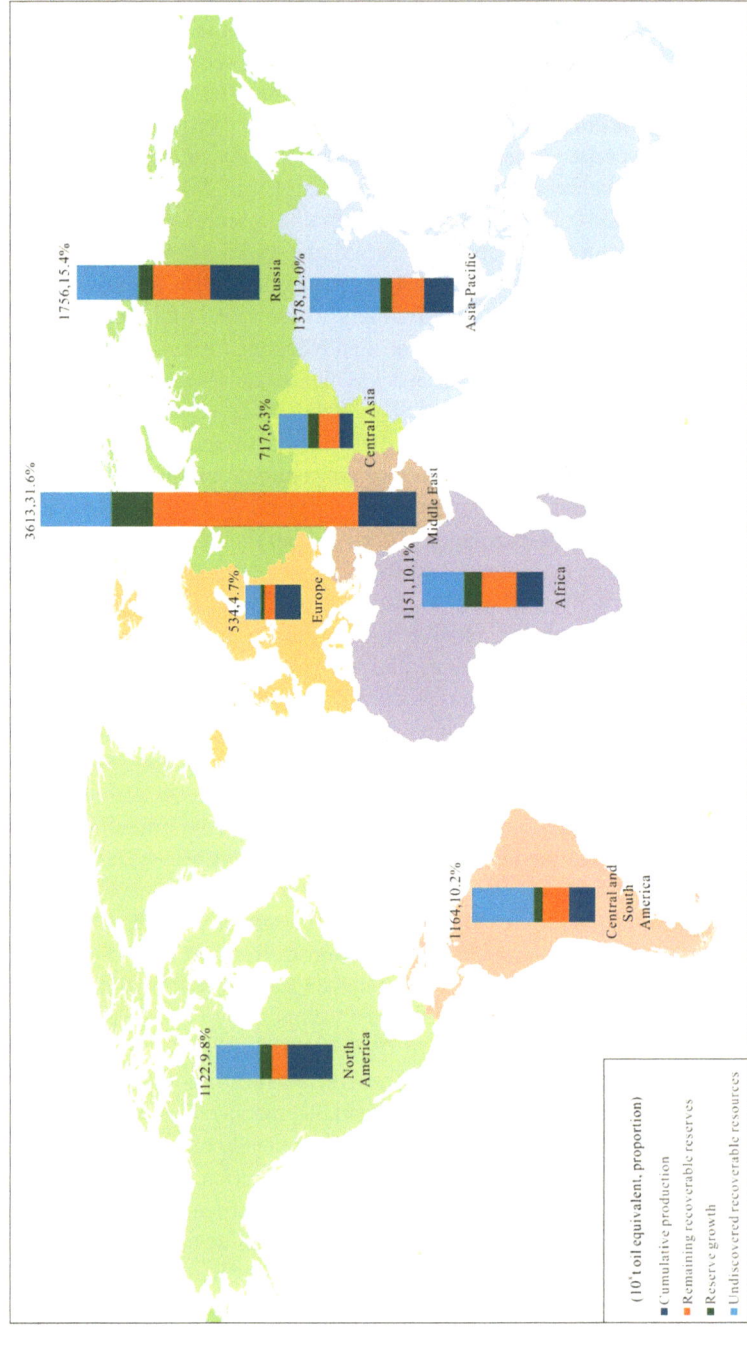

Fig. 1 Planar distribution of conventional oil and gas resources in different regions worldwide (Miller Cylindrical Projection)

Table 1 Statistics on regions with various types of conventional oil and gas resources worldwide

Region	Cumulative production			Remaining recoverable reserves			Reserve growth from discovered oil and gas fields			Undiscovered recoverable resources			Total (10^8 t oil equivalent)
	Oil (10^8 t)	Condensate (10^8 t)	Gas (10^{12} m^3)	Oil (10^8 t)	Condensate (10^8 t)	Gas (10^{12} m^3)	Oil (10^8 t)	Condensate (10^8 t)	Gas (10^{12} m^3)	Oil (10^8 t)	Condensate (10^8 t)	Gas (10^{12} m^3)	
North America	223.48	11.08	22.86	89.48	5.26	7.18	69.77	4.07	4.23	160.08	56.01	24.58	1122.2
Central and South America	179.12	3.80	7.39	153.27	10.59	10.51	51.09	2.64	2.78	435.50	20.58	15.33	1164.3
Europe	105.72	6.85	15.61	38.39	5.27	7.07	8.79	2.75	2.43	56.45	13.68	9.56	534.3
Africa	179.58	8.72	6.93	122.20	19.96	22.85	76.06	5.20	9.87	157.07	47.44	22.91	1151.0
Middle East	489.90	14.29	6.02	911.81	105.93	112.53	167.08	34.70	22.88	296.03	47.47	39.41	3612.9
Central Asia	51.69	9.84	8.45	47.10	8.07	17.38	20.59	5.51	8.18	54.99	12.78	25.20	716.7
Russia	261.87	0.91	24.11	194.63	24.21	39.99	63.79	3.72	7.60	148.84	38.43	47.58	1755.9
Asia-Pacific	159.47	9.85	12.97	73.26	15.17	26.25	24.65	4.91	9.22	221.05	21.71	50.72	1377.6
Total	1650.8	65.3	104.3	1630.1	194.5	243.8	481.8	63.5	67.2	1530.0	258.1	235.3	11434.9

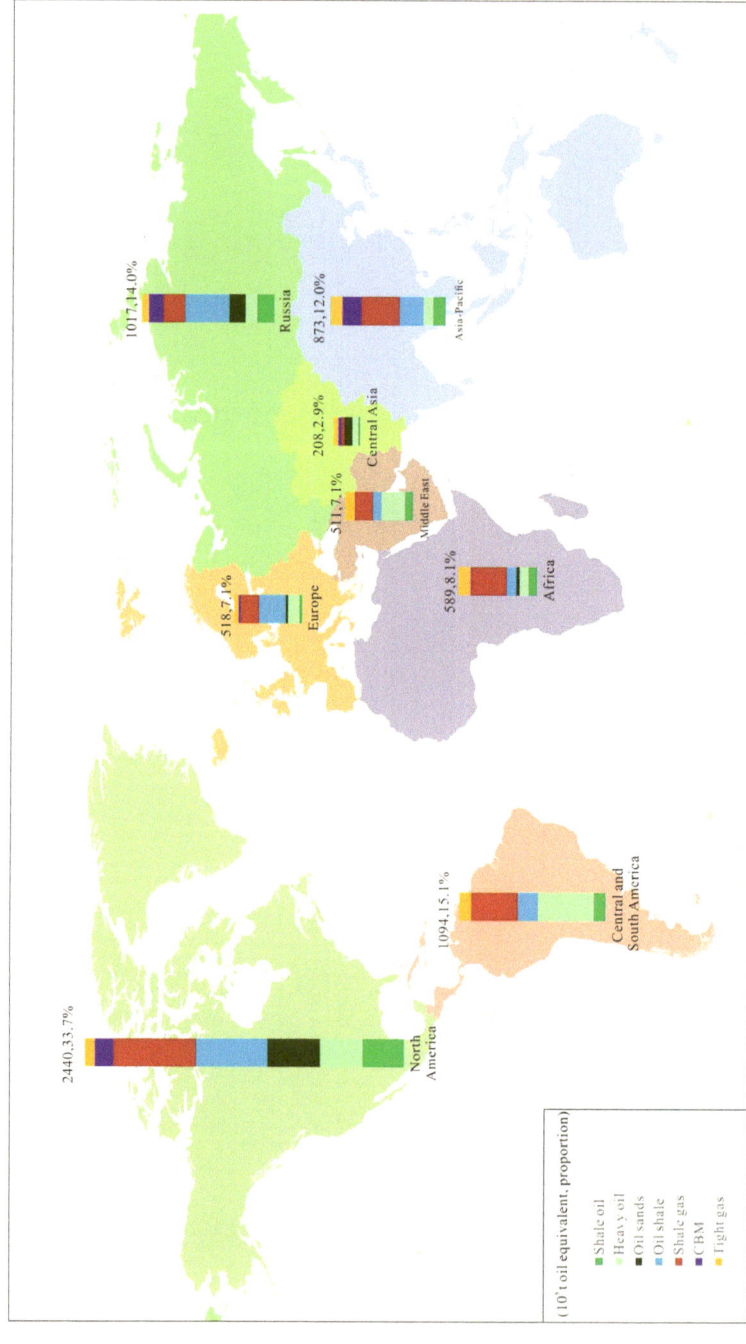

Fig. 2 Planar distribution of unconventional oil and gas resources in various regions worldwide (Miller Cylindrical Projection)

Table 2 Statistics on regions with different types of unconventional oil and gas resources worldwide

Region	Technically recoverable unconventional oil resources (10^8 t)				Technically recoverable unconventional gas resources (10^{12} m^3)			Total (10^8 t oil equivalent)
	Shale oil	Heavy oil	Oil sands	Oil shale	Shale gas	CBM	Tight gas	
North America	313.59	324.70	403.39	544.53	74.32	16.98	8.60	2440.1
Central and South America	89.38	418.20	0.00	153.23	40.48	0.03	10.20	1094.2
Europe	23.63	84.24	17.93	200.09	16.69	2.01	3.73	517.6
Africa	63.35	64.77	25.00	69.68	31.66	0.59	10.60	589.0
Middle East	59.00	180.59	0.00	62.65	16.08	0.00	8.40	511.5
Central Asia	16.61	44.57	59.40	0.00	2.68	3.20	4.30	207.6
Russia	130.26	88.96	125.67	338.18	19.66	13.11	6.34	1017.3
Asia-Pacific	94.60	68.76	7.52	172.42	34.30	17.13	10.49	872.5
Total	790.4	1274.8	638.9	1540.8	235.9	53.0	62.7	7249.9

69.7% of the global recoverable unconventional oil resources are distributed in North America, Russia, and Central and South America. The recoverable unconventional oil resources in the top five countries including the United States, Russia, Canada, Venezuela, and Saudi Arabia, account for more than 63.2% of the global recoverable unconventional oil resources. 75% of the global recoverable unconventional oil resources are concentrated in the top 20 basins, including the Alberta basin, Eastern Venezuela basin, Arabian basin, and others. The global recoverable unconventional oil resources are mainly distributed in three types of basins, namely foreland basins, cratonic basins, and rift basins, and are concentrated in the Cretaceous, Jurassic, Paleogene, Neogene, Devonian, and other formations. The technically recoverable shale oil (including tight oil) resources are 790.4×10^8 tons, accounting for 18.6% of the global recoverable unconventional oil resources; the technically recoverable heavy oil resources are 1274.8×10^8 tons, accounting for 30.0% of the global recoverable unconventional oil resources; the technically recoverable oil sands are 638.9×10^8 tons, accounting for 15.1% of the global recoverable unconventional oil resources; and the technically recoverable oil shale resources are 1540.8×10^8 tons, accounting for 36.3% of the global recoverable unconventional oil resources.

60.4% of the global recoverable unconventional gas resources are concentrated in North America, Asia-Pacific, and Central and South America. The recoverable unconventional gas resources in the top five countries, including the United States, Russia, China, Canada, and Argentina, account for more than 55.9% of the global recoverable unconventional gas resources. 80% of the global recoverable unconventional gas resources are concentrated in 26 basins, such as the Alberta Basin, Gulf Coast Basin, Zagros Basin, Appalachian Basin, Neuquén Basin, and others. The global recoverable unconventional gas resources are mainly distributed in three types of basins: foreland basins, cratonic basins, and rift basins. These resources are concentrated in the Cretaceous, Jurassic, Carboniferous, Devonian, and Silurian formations. The technically recoverable shale gas resources are 235.9×10^{12} m^3, accounting for 67.1% of the global recoverable unconventional gas resources. The technically recoverable CBM resources are 53×10^{12} m^3, representing 15.1% of the global recoverable unconventional gas resources. Additionally, the technically recoverable tight gas resources are 62.7×10^{12} m^3, making up 17.8% of the global recoverable unconventional gas resources.

By analyzing the geological conditions and potential of undiscovered oil and gas resources, it has been found that the key exploration areas for conventional oil resources in the future are mainly distributed in seven major groups of basins. These include basins along the passive continental margin of the North Atlantic, basins along the passive continental margin of the Atlantic, Russian foreland-rift basins, and Zagros/Arabian foreland basins. The key exploration areas for conventional gas resources in the future are mainly distributed in six major groups of basins, namely Zagros/Arabian foreland basins, basins along the passive continental margin of the offshore East Africa, basins along the passive continental margin of the key favorable exploration areas for shale oil resources in the future are mainly distributed in seven major groups of basins, including those in the North Atlantic, Central Asian rift-foreland basins, Russian foreland-rift basins, and basins along the passive continental margin of the Arctic. These include North American foreland basins, Andean foreland basins, North African cratonic basins, Central and West African rift basins, Russian foreland-rift basins, Zagros/Arabian foreland basins, and Southeast Asian back-arc basins. The key exploration areas for shale gas resources in the future are mainly distributed in seven major groups of basins, including North American foreland basins, Central and South American cratonic basins, Andean foreland basins, North African cratonic basins, Russian foreland-rift basins, Zagros/Arabian foreland basins, and Central Australian cratonic basins.

Considering the progress in exploration, the tendering plans of resource-rich countries, and the conducive environments for cooperation, it is recommended to focus on the following ten major opportunities for cooperation in the near future: the coastal basins in Guyana and surrounding areas, the periphery of the Caribbean Sea, the pre-salt formations on both sides of the South Atlantic, South African offshore, the Argentine offshore, Northwest African offshore, the East African offshore basins, deep plays in the Zagros Basin, the Russian Arctic, and the East African Rift System.

Contents

Chapter 1
History and Outlook of Evaluation on Global Oil and Gas Resources

Oil and natural gas are the most important primary energy sources in the world. Many domestic and foreign institutions predict that even if China puts forward the goal of "Carbon Peaking and Carbon Neutrality", China's oil and gas demand in 2050 will reach or exceed 3×10^8 t and 5000×10^8 m^3 (RIPED 2021; BP 2021). Considering that, the rational and effective use of overseas oil and gas resources will remain one of the important strategies for ensuring China's national energy security and promoting low-carbon development in the next 30 years. It is the main content of global oil and gas resource evaluation to scientifically predict the distribution of remaining recoverable oil and gas reserves and the undiscovered resources, and to reasonably evaluate the reserves growth potential of discovered oil and gas fields in the world, which is also the basis for the international oil companies to implement the internationalization strategies.

Global oil and gas resource evaluation has a history spanning a hundred of years. White first reported the global oil and gas resource quantity in 1920 (White 1920), and RIPED, CNPC released the results of global oil and gas resource evaluation completed in 2020 (RIPED 2021). Since the start of oil and gas resource evaluation, scholars, such as Pratt (1942), Levorsen (1950), Weeks (1948, 1958, 1965, 1971) and Adams et al. (1975), and institutions, including Exxon (1976), Nehring (1979), Martin (1985), Masters et al. (1987, 1991, 1994, 1997), Campbell (1989, 1992, 1997), Odell (1998), Cerigaz (2001), Schmoker (2002), and the United States Geological Survey (USGS 2003, 2005), carried out a large number of effective research work. Since 2008, USGS (2012) and PIPED (RIPED 2021; Tong 2014, 2018) have been the main institutions that carried out the global oil and gas resource evaluation and released the evaluation results. Because the evaluation process and results are affected by many factors, the exact data of global oil and gas resource evaluation has not been recognized so far. The analysis shows that there are two kinds of understanding about the current evaluation results: (1) The world is rich in oil and gas resources, which should be actively explored. As long as new resources are discovered through the exploration, the oil and gas reserves will continue to increase. (2) According to the

© The Author(s) 2024

L. Dou et al., *Global Oil and Gas Resources: Potential and Distribution*,
https://doi.org/10.1007/978-981-97-4756-6_1

oil and gas depletion theory, the oil and gas resources will soon be depleted through the development. Based on comprehensive analysis of the process, methods and fields of global oil and gas resource evaluation, the latest global oil and gas resource evaluation results released by RIPED were systematically analyzed and compared with the evaluation results published by the major international research institutions. In addition, the main influencing factors of the global oil and gas resource evaluation results were discussed in combination with the continuous progress of oil and gas geological theory and exploration and development technology. To provide reference and guidance for global oil and gas resource evaluation in the future and provide reference and basis for oil companies to implement internationalization strategy, the development trend was studied.

1.1 History and Stage of Evaluation on Global Oil and Gas Resources

According to the incomplete statistics, since the 1920s, about 50 institutions and scholars have carried out the global oil and gas resource evaluation (RIPED 2021; USGS 2012) (Fig. 1.1). Among them, Weeks (1948, 1958, 1959, 1965, 1971) and Campbell (1989, 1992, 1997) both published the evaluation results for seven times, which promoted the development of global oil and gas resource evaluation. The statistics show that the resource evaluation results in different ages vary greatly, up to more than 100 times, which is closely related to the geological understanding, discovery of oil and gas fields, exploration progress and efficiency, geographical scope of evaluation and evaluation method adopted in different periods (Guo 2015, 2016; Zheng 2019; Wang 2021; Tao 2015), as well as the international oil price and crude oil production at the time of evaluation. The development process of global oil and gas Resource evaluation can be divided into four main stages considering resource evaluation methods, exploration technologies, geological theories, and the types of traps discovered.

1.1.1 Initial Stage (1900–1957)

The birth and initial stage of global oil and gas resource evaluation mainly benefited from the fact that USA entered the peak period of oil and gas discovery and accumulated many systematic oil and gas reserves and production data. At this stage, the global petroleum industry was in its infancy, the main exploration object was the anticline structure, and the theory of petroleum geology was still forming. During this period, few institutions carried out global oil and gas resource evaluation. Scholars, represented by Weeks (1948) of Standard Oil, used the sedimentary rock volumetric production method based on the geological analysis to evaluate the global oil and

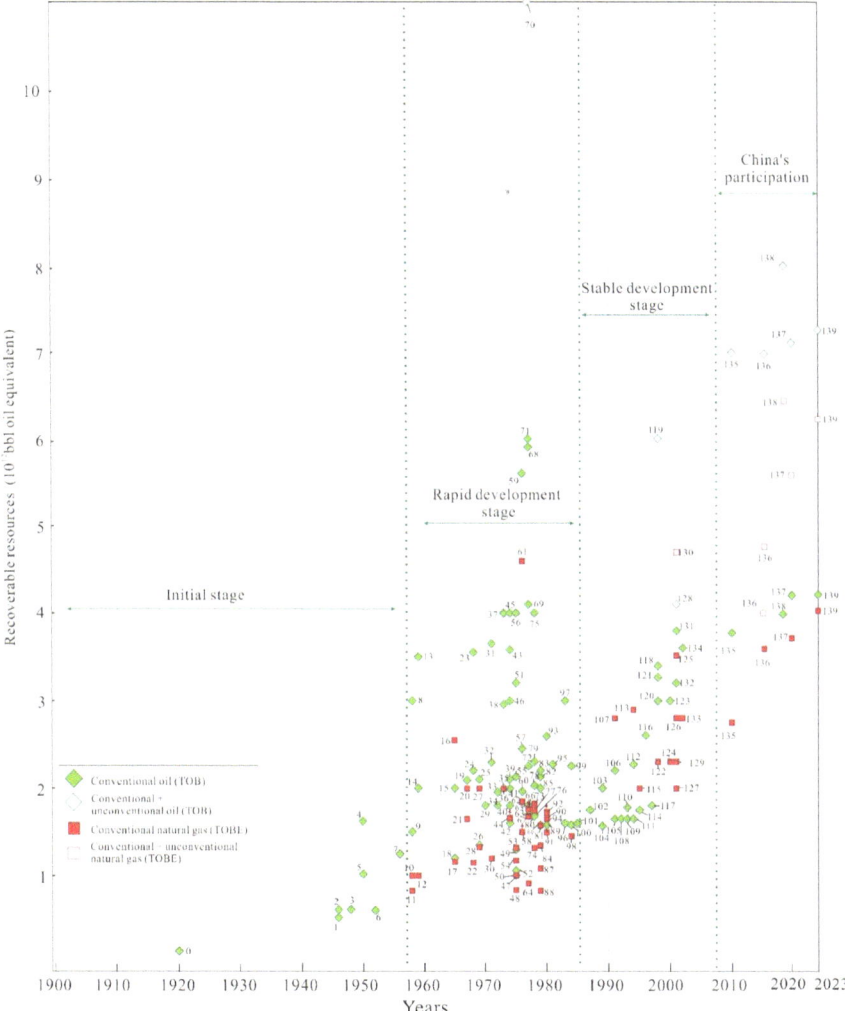

Fig. 1.1 Global oil and gas resources evaluation and phase division. *Note* Data 0 comes from Reference (White 1920), data 1–134 comes from Reference (USGS 2005), data 135 comes from Reference (Tong 2014), data 136 comes from Reference (Tong 2018), data 137 comes from Reference (RIPED 2021,excluding China data), and data 138 comes from Reference (Shell 2001).data 139 comes from the current dataset (China data come from Zheng 2019; Wu 2022; Wang 2023)

gas resources. The method was essentially an area/volume analogy method, using the sedimentary rock per cubic mile or the surface per square mile as the assessment unit. The ultimate recoverable reserves (URR) of a basin were obtained by assessing its crude oil or natural gas production and applying the assessment to the basin. The final recoverable global oil resources predicted at this stage range from 0.063 to 1.800TBO (10^{12} bbl), and the natural gas resource potential is not predicted

(Fig. 1.1). Overall, the evaluation scope of the initial stage mainly covers the land, with the limited sedimentary basins evaluated.

1.1.2 Rapid Development Stage (1958–1985)

During this period, the discovery of oil and gas in USA entered a low tide, and its dependence on foreign oil and gas continued to rise, while a series of large and giant oil and gas fields were discovered in the Middle East, Northern Europe, Southeast Asia, the Soviet Union, North Africa onshore and many offshore shallow waters worldwide. Meanwhile, petroleum geological theories were rapidly developed and improved, opening the rapid development stage in terms of the evaluation of global oil and gas resources. At this stage, different evaluation methods were developed accordingly. American scholars made extrapolations based on the discovery history of oil and gas fields, discovery rate of oil and gas fields, drilling footage, oil and gas production rate and the size of known oil fields, and proposed the statistical methods such as the growth curve prediction model (Weeks 1950; Hubbert 1969) and the oilfield size sequence (Hubbert 1974; Kaufman 1965), while the Soviet scholars actively explored the application of geochemical material balance method (genetic method) in oil and gas resource evaluation. The final recoverable global oil and gas resources predicted were predicted to be 1–6 TBO and 0.8–4.5 TBOE (10^{12} bbl oil equivalent), respectively. Among them, the final recoverable resources of global oil estimated by the statistical methods were 2–3 TBO. Styrikovich, a Soviet scholar, predicted that the final recoverable resources of global oil (including the unconventional oil) were 11 TBO by adopting the genetic method (Styrikovich 1977) (Fig. 1.1). At this stage, the global oil and gas resource evaluation began to cover the offshore shallow waters, and a preliminary evaluation of unconventional oil and gas resources such as coalbed methane and oil shale was carried out.

1.1.3 Stable Development Stage (1986–2007)

During this period, most of the world's onshore sedimentary basins and offshore shallow water areas had undergone at least one round of oil and gas exploration, numerous discoveries were made in oil and gas exploration, and related study reports also increased greatly, providing more reliable geological data for global in-depth oil and gas resource evaluation. At this stage, the oil and gas resource evaluation methods were continuously developed and optimized, and the assessment units were more rationalized. With the application of statistical methods, the idea that the samples of assessment units shall be of statistical significance had been increasingly recognized. The independent petroleum system (IPS) and the total petroleum system (TPS) became widely adopted assessment units by scholars and institutions (USGS 2000, 2005). Meanwhile, thanks to the development of computer technology, the genetic method had gradually developed into the basin simulation method (Wu 2005; Long 2005; Zhao 2019). USGS researchers, represented by Masters, began to continuously release the evaluation results of global oil and gas resources, and subdividing them into different categories, including the production, remaining recoverable reserves, reserves growth and resources to be discovered (Masters 1987, 1991, 1994, 1997). At this stage, the oil and gas resources evaluation method, evaluation process and evaluation results all tended to be stable, with the total recoverable oil resources in the world ranged; from 1.5 to 3.0 TBO and the recoverable natural gas resources ranging from 1.0 to 2.0 TBOE (Fig. 1.1).

1.1.4 China's Participation Stage (2008–Present)

In the early 1980s, China began to carry out domestic oil and gas resource evaluation on a large scale, and the relevant departments had organized and carried out several rounds of oil and gas resource evaluation in domestic oil and gas basins (Guo 2016; Zheng 2019), which has accumulated methods, software and talents for Chinese scholars and institutions to conduct global oil and gas resource evaluation. In 1993, CNPC began to go abroad to carry out oil and gas exploration and development. Up to now, CNPC has explored more than 50 oil and gas basins in 5 continents, which not only discovered many oil and gas fields, but also accumulated rich geological data, laying a solid data foundation for global oil and gas resource evaluation.

Since 2008, relying on national major science and technology projects such as "Large-scale Oil and Gas Fields and Coalbed Methane Development", CNPC has innovated and developed the evaluation methods for the conventional and unconventional oil and gas resources with "play" as the unit, comprehensively completed the geological and resource potential evaluation of conventional oil and gas resources and 7 types of unconventional oil and gas resources in the major oil and gas basins in the world, obtaining the evaluation data with independent intellectual property rights. The evaluation results were released twice in 2017 and 2021, attracting wide attention and application in the society (Tong 2018; RIPED 2021). According to the results of the third round of independent evaluation by RIPED, the total recoverable resources of conventional oil and natural gas in the world are 4.28 TBO and 4.06 TBOE respectively. The total recoverable resources of unconventional oil and gas are 3.1 TBO and 2.19 TBOE, respectively (overseas data according to RIPED 2021; China data according to Zheng 2019; Wu 2022; Wang 2023). At this stage, USGS, the main foreign institution conducting global oil and gas resource evaluation, continuously updated the evaluation results based on the global oil and gas resource evaluation results released in 2000 and published the latest global oil and gas resource evaluation results in 2012 (USGS 2012).

1.2 Main Influencing Factors of Resource Evaluation Results

By reviewing the development history of global oil and gas resource evaluation, the analysis shows that the evaluation method, oil and gas geological theory, oil and gas exploration technology, evaluation field and scope, oil price and production of oil and gas have important effects on the evaluation results.

1.2.1 Evaluation Method

Since the study of global oil and gas resources evaluation development of the evaluation methods involved can be summarized as the analogy method, statistical method, genetic method, Delphi method and their combinations. Different evaluation methods have their own advantages and disadvantages in terms of applicability and evaluation effect (Hu 1992; Rice 1992; Long 2005; Wu 2005; Zhang 2006; Tao 2015; Guo 2016; Zhao 2019; Zheng 2019; Wang 2021).

1.2.1.1 Analogy Method

The analogy method is mainly suitable for the basins with low exploration maturity, which requires less data, but the accuracy of its evaluation results is relatively low. The analogy method is based on the main geological elements of reservoir formation, and the evaluation results depend on the similarity between the basin and the analogue object and the evaluator's identification of all trap types in the basin (Guo 2015, 2016). For the basins with different resource types and exploration maturity, different analogy methods can be used, such as the resource abundance analogy method and the estimated ultimate recovery (EUR) analogy method. In the 4th round of national oil and gas resource evaluation organized by CNPC from 2010 to 2015, the database of 203 scale areas was established, providing the analog data basis for the application of analogy method.

1.2.1.2 Statistical Method

With the continuous advancement of oil and gas exploration, American scholars, represented by Kaufman, began to use the statistical methods to predict the future growth potential of oil and gas resources (Kaufman 1965; Hu 1992; Rice 1992; Wu 2005). This method is suitable for the basins with medium and high exploration maturity and requires a high degree of data detail and statistical significance of the samples. USGS has been using the statistical methods to carry out global oil and gas resource evaluation, and has continuously optimized the statistical models, developing from truncated and tailed Pareto simulation method to the seventh approximation model (USGS 2003, 2005, 2012). However, because one specific method is adopted for the assessment units with different exploration maturity, the applicability of the statistical method is relatively poor.

1.2.1.3 Genetic Method

With the rise of organic geochemistry, especially the wide application of basin simulation method, the genetic method has been widely promoted. Based on the characteristics of source rocks, this method is used to simulate the process of oil and gas generation and expulsion by the computer, to obtain the quantities of oil and gas generation, oil and gas expulsion, oil and gas loss and accumulation, requiring high comprehensiveness of data (Wu 2005; Guo 2015; Zhao 2019). With the genetic method, the upper limit of oil and gas resources in the basin can be calculated, but it is difficult to accurately estimate the migration and accumulation coefficient (efficiency) of oil and gas. Hence, its applicability is relatively poor.

1.2.1.4 Delphi Method

By inviting experts to score the distribution parameters of oil and gas resources in a basin, the Delphi method can be used to quickly determine the oil and gas resource potential of a basin (Hu 1992; Rice 1992; Long 2005; Zhang 2006; Tao 2015; Guo 2016). The method is simple and feasible. It is mainly applicable to the early exploration stage of the basin. The reliability of the evaluation results depends on the invited experts' understanding of the relevant basin and their evaluation level.

For the unconventional oil and gas resources, due to their diverse types, control by source rocks and tight reservoirs, and continuous accumulation and distribution, the evaluation methods adopted for different resource types are quite different. The volumetric method and analogy method are more commonly used. Among them, the analogy method is mainly based on the EUR analogy method of producing wells, such as the FORSPAN model method adopted by USGS (2003). The volume method is mainly used to calculate the quantity of oil and gas resources by calculating the abundance of unconventional oil and gas resources (such as gas content and oil content, etc.) in the geological body per unit volume. Genetic method is also commonly used to evaluate the shale oil and gas resource potential controlled by source rocks (Zou 2011). By calculating the oil and gas generation potential of source rock systems, this method can be used to further calculate the quantity of residual oil and gas in source rocks, thereby obtaining the quantity of the unconventional oil and gas resources.

1.2.2 Oil and Gas Geological Theory

Oil and gas resource evaluation is to predict the oil and gas resources under the guidance of geological models and geological understanding. So, the continuous improvement of oil and gas geological theory will affect the scope and results of global oil and gas resource evaluation, making the evaluation results more reliable, practical and instructive.

When oil was first drilled in Pennsylvania in 1859, the anticlinal theory of the mid-nineteenth century was the germ of modern petroleum geology (White 1885). In 1934, the trap theory was proposed (McCollough 1934). In 1950, Turbidity current theory was published (Kuenen 1950). In 1956, Petroleum Geology written by Levorsen, an American scholar, was published (Levorsen 1956), marking the gradual systematization of modern petroleum geology theory. The oil and gas geological theory and exploration in this period did not cover the lithology, formation, and subtle traps, resulting in the underestimation of the most oil and gas resource potentials (Weeks 1950, 1958, 1959).

The early 1960s to the late 1970s was the peak stage of oil and gas exploration and discovery in the world, and it was also the stage of rapid development of petroleum geological theories. Important theories such as kerogen oil and gas generation theory, plate tectonics theory, basin type theory, carbonate reservoir theory and secondary migration and accumulation theory of oil and gas were formed (Wilson 1963; Mckenzie 1969; Dickinson 1974, 1976; Klemme 1988; Bally 1980). In the 1970s, Halbouty, an American petroleum geologist, put forward the concept of "subtle reservoirs" and expanded the field of oil and gas exploration to include stratigraphic traps, lithologic traps and many types of complex reservoirs (Halbouty 1970). The theory of kerogen oil and gas generation strongly supports the theory of organic oil generation (Tissot 1978). In this period, seismic stratigraphy, sequence stratigraphy, underwater fan, petroleum system, debris flow and other theories came into being (Vail 1977; Walker 1978; Margoon 1994; Shanmugam 1996; Cross 1998), wIn this period, seismic stratigraphy, sequence stratigraphy, underwater fan, petroleum system, debris flow and other theories camehich greatly enriched the contents and methods of petroleum geology study. Thanks to the birth and guidance of these petroleum geological theories, the global oil and gas resource evaluation has entered a stage of rapid development, and many institutions and scholars have carried out oil and gas resource evaluation and released evaluation results. Because the evaluation object, field and scope are obviously expanded, the result of the oil and gas resource evaluation in this period is also more optimistic.

In the twenty-first century, under the guidance of new theories and the advancement of new technologies, major breakthroughs have been made in the global unconventional oil and gas exploration and development, and the oil sands, heavy oil, tight gas, and coalbed methane have become the key areas of unconventional oil and gas development (Halbouty 1979; Meimer 2007; Jarvie 2012; Zou 2011, 2020). Among them, shale gas has become the hot-spot direction of unconventional natural gas, and shale oil has become the "bright spot" of unconventional oil development. During this period, unconventional oil and gas resources became an indispensable part of global oil and gas resource evaluation.

1.2.3   Oil and Gas Exploration Technology

While the reserves of oil and gas resources in the world are a constant, human understanding of global oil and gas resources is deepening with the progress of science and technology and the continuous development of exploration technology. During the first oil crisis of the 1970s, the "oil depletion theory" was prevalent, considering that the oil industry would soon reach its end (Hubbert 1956, 1974; Mccabe 1997). However, with the progress of technology, the quantity of oil and gas resources that can be discovered and developed by mankind is not less and less, but more and more.

The development of exploration technology makes the objects of oil and gas resources evaluation more and more abundant, directly increasing the total quantity of oil and gas resources evaluation. With the development of seismic data acquisition, processing, and interpretation technology, it is possible to understand deeper strata, finer structures, and find subtle oil and gas traps. The development of drilling technology has not only improved the speed of oil and gas discovery and greatly accelerated the development process of the petroleum industry. More importantly, it has enabled geologists to discover high-quality condensate and natural gas resources in the deep and ultra-deep plays that were impossible to reach previously, so that the occurrence of oil and gas resources has expanded to deep layer. Advances in seismic and drilling technology have enabled the geologists to move toward a "deep ocean" understanding of oil and gas resources, achieving the exploration and utilization of oil and gas resources in deepwater areas of more than 3000 m (Fig. 1.2) (Zou 2011; Walker 1978; Shanmugam 1996).

In addition, some exploration technologies can also be used as input parameters to directly participate in the evaluation of oil and gas resources. For example, exploration efficiency is introduced in the discovery process method, to adjust the influence of exploration technology on the evaluation results, and the scale of resources discovered by drilling progress trend method is directly used as the input parameters. With the development of exploration technology, the evaluation method based on the evaluation parameter will optimize the prediction of oil and gas resources and make the evaluation result more scientific and objective.

1.2.4   Evaluation Domain and Scope

With the progress and development of oil and gas geological theory and exploration technology, the leapfrog development has been realized from the initial structural oil and gas reservoir to stratigraphic oil and gas reservoir, composite oil and gas reservoir,

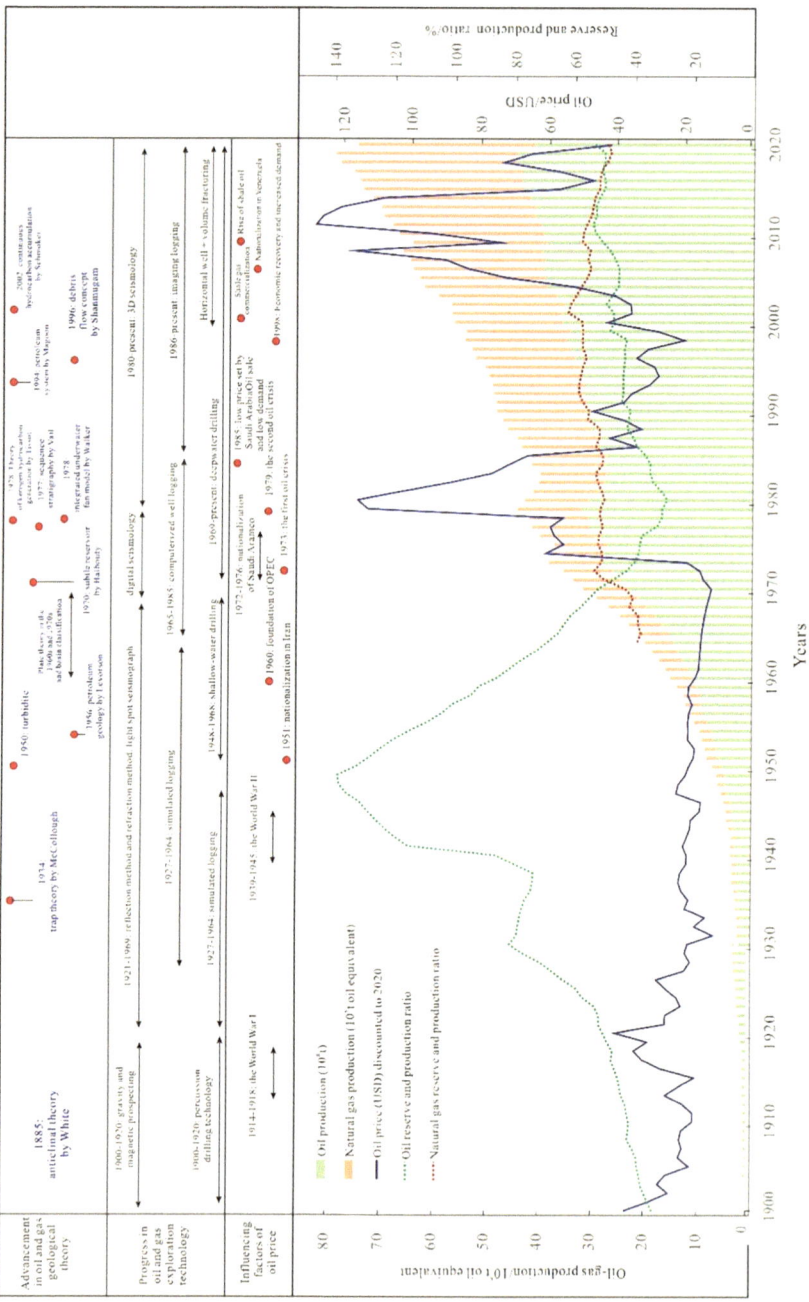

Fig. 1.2 Coupling relationship between geological theories, exploration and development technologies, oil prices and production

continuous oil and gas reservoir, from land to ocean, from shallow water area to deep water area, and from conventional oil and gas reservoir to unconventional oil and gas reservoir. The types, numbers and recoverable reserves of the newly discovered traps in different years of the world (Fig. 1.3), the number of oil and gas fields and the onshore and offshore distribution of reserves (Fig. 1.4) all indicate that the field and scope of global oil and gas resource evaluation are gradually expanding.

Before 1958, oil and gas discoveries were mainly concentrated on the land. The scope of oil and gas resource evaluation was limited to land (Weeks 1948). With the advancement of exploration technology, more and more oil and gas discoveries have been made in shallow water with a depth of less than 500 m. By 1985, oil and gas were discovered in deep water with a depth of more than 500 m. After 1996, it entered the stage of oil and gas discovery in the ultra-deep water (the depth greater than 1500 m). With the expansion of offshore oil and gas exploration and discovery, the geographical scope of oil and gas resource evaluation is also expanding. After 1958, the United States began to carry out the evaluation of oil and gas resources undiscovered in the offshore area, respectively evaluating the resource potentials of the water depths of 200 m (660 ft), 457 m (1500 ft), 1000 m (3300 ft), 1830 m (6000 ft) and 2500 m (8200 ft) (USGS 2012).

Statistics of global oil and gas resource evaluation results released by different generations, different stages, different scholars, or institutions show that in the 149 sets of evaluation data, 110 sets are oil evaluation data, of which 7 sets are unconventional oil evaluation results. 39 sets are gas evaluation data, of which 5 sets include the unconventional natural gas resources (USGS 2005). The early global oil and gas resource evaluation only focused on the conventional oil and gas resources. It was not until 1977 that Soviet scholar Styrikovich first provided the evaluation results of unconventional oil resources (Styrikovich 1977). In 2001, Shell Oil Company released the evaluation results of unconventional gas resources for the first time (Shell 2001). There are various types of unconventional oil and gas resources, from the early visible resources such as heavy oil and oil sands to the emerging shale oil, tight oil and shale gas in recent years. The types and quantities of unconventional resources evaluated at different stages are bound to be very different.

1.2.5 Oil Price and Oil/Gas Production

Oil and gas resource evaluation is a prediction way based on scientific method, which is largely influenced by the evaluator's expectation of future economic trend. Oil price fluctuations and oil and gas production will affect the expectation of oil and gas resource evaluation results to a certain extent (Mccabe 1997). When oil and gas production and oil prices are high, different scholars and institutions are optimistic about the future economic upswing. Therefore, the value of resource evaluation parameters and evaluation results are also optimistic, and vice versa.

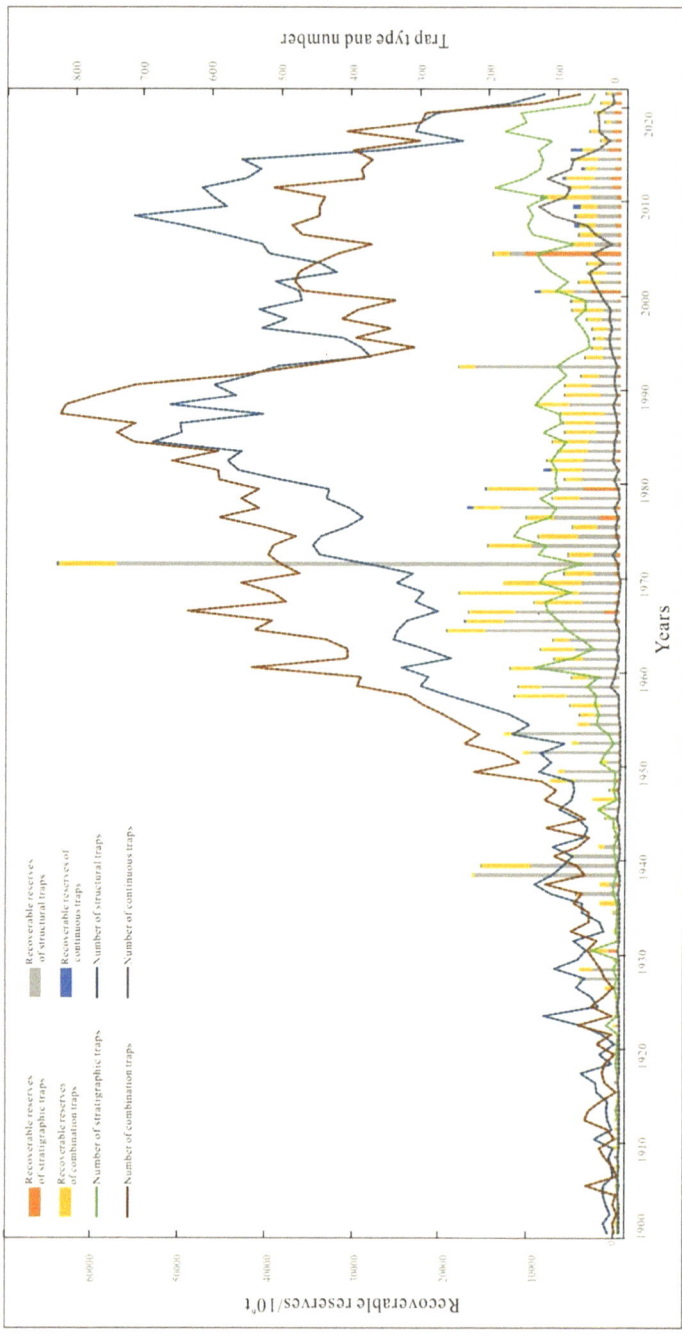

Fig. 1.3 Numbers and discovered reserves of traps in different types (data from IEA 2021, excluding the land data of North America)

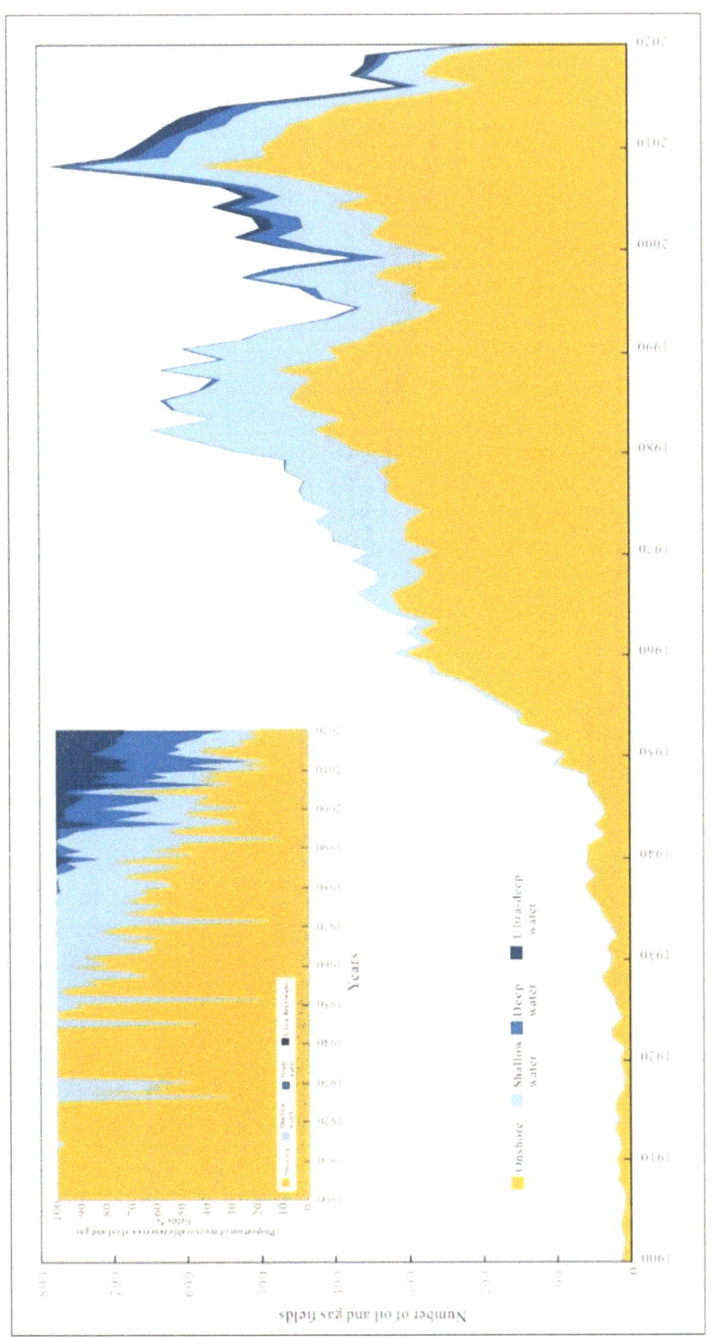

Fig. 1.4 Onshore-offshore distribution of global oil and gas reserves discovered in different years (data from S&P Global 2021), excluding the land data of North America)

1.3 Evaluation Methods and Results

1.3.1 Evaluation Method

Since 2008, RIPED has completed three rounds of global oil and gas resource evaluation. In the evaluation process, CNPC took the basin as an integral part, started from the effective source rocks, comprehensively considered multiple types of parameters, and established the evaluation process of conventional and unconventional oil and gas resource types, to ensure that important resources in the basin would not be missed or double-counted. Moreover, RIPED developed a suitable resource evaluation methods based on the exploration maturity of the assessment unit. For the potential evaluation of conventional oil and gas resources to be discovered, the statistical method based on discovery process method was used in mature basins, the subjective probability method was used in basins with medium exploration mature basin, and the analogy method based on geological evaluation was used in frontier basins. The method of 11 reserves growth curves established for different regions was used to predict the reserves growth of discovered oil and gas fields. The parameter probability method, GIS spatial graph interpolation method, genetic constraint volume method and hyperbolic exponential decline method were used to evaluate the recoverable resources, for which the unconventional oil and gas technologies were adopted. Finally, a complete conventional and unconventional oil and gas resource evaluation system based on database and network environment was established, to ensure the reliability and uniformity of evaluation results (Table 1.1) (Tong 2014, 2018; RIPED 2021).

During the period of 2016–2023, RIPED made scientific evaluation and prediction on the total oil and gas resources of 468 basins in the world in the next 30 years through in-depth and systematic global oil and gas geological evaluation, including the remaining recoverable reserves of 34,119 oil and gas fields and their growth potential and distribution, the potential and distribution of recoverable resources undiscovered in 920 conventional oil–gas plays, the potential and distribution of technically recoverable resources in 656 unconventional oil and gas plays of 7 types (shale oil/tight oil, oil shale, heavy oil, oil sands, tight gas, shale gas and coalbed methane) (RIPED 2021). In order to compare the evaluation results published by other institutions with the same criteria, China's oil and gas evaluation data is selected from the fourth round of resource evaluation results released by CNPC in 2019 (excluding the reserves growth of discovered oil and gas fields and the evaluation results of condensate oil) (Zheng 2019; Wu 2022; Wang 2023).

Table 1.1 Comparison of global oil and gas resources evaluation methods between RIPED and USGS

Comparison items	USGS	RIPED	Conclusion
Assessment unit	Assessment unit (AU) and total petroleum system (TPS) as main units	Play as the main assessment unit	The play is the object of oil and gas exploration. The evaluation process of RIPED is more specific, and the evaluation results can be used to directly guide exploration and production as the play is used as the assessment unit
Method diversity and applicability	Regardless of the exploration maturity, a single method is used Conventional oil and gas: the seventh approximation model, etc Unconventional oil and gas: based on the producing well productivity method	According to the maturity of exploration, appropriate methods are adopted Conventional oil and gas: statistical method, subjective probability method, and analogy method Unconventional oil and gas: parameter probability distribution volume method, GIS spatial graph interpolation method, genetic constraint volume method, and hyperbolic exponential decline method	USGS: The method is simple and the difference in data acquisition and exploration maturity is not considered RIPED: Consider the acquisition of data and the maturity of exploration and choose the appropriate method
Database support and result management	Completed in Excel; there was no dedicated database in the early stage, but it was supported by the database during the evaluation in 2012	Directly based on resource evaluation project database and project management platform; based on the global oil and gas exploration and development database	USGS: There is no special database and data bank RIPED: The basic database provides the basic evaluation data, and the project database is used to manage all evaluation parameters and results
Application software system	In the Excel, the business tools (crystal ball) are embedded, and no special application software system is provided	A complete and web-based application system for conventional and unconventional oil and gas resource evaluation has been established and is jointly managed by the global oil and gas resource evaluation working platform	USGS: No complete application software system is developed RIPED has established a complete evaluation system for the conventional and unconventional oil and gas resources based on database and network environment

1.3.2 Conventional Oil and Gas Resources Evaluation Results

The recoverable resources of conventional oil and gas consist of four parts: the cumulative production, remaining recoverable reserves, reserves growth of discovered oil and gas fields and recoverable resources undiscovered. According to the evaluation results of RIPED in 2023, the recoverable resources of global conventional oil, condensate oil and conventional natural gas are $5,292.8 \times 10^8$ t, 581.9×10^8 t and 650.6×10^{12} m^3, respectively. The cumulative production of conventional oil and gas is 2608×10^8 t oil equivalent, and the recovery rate is 22.8%. The remaining recoverable reserves of conventional oil and gas, the reserves growth of discovered oil and gas fields and the recoverable resources of conventional oil and gas undiscovered are 3908.1×10^8 t oil equivalent, 1119.6×10^8 t oil equivalent and 3299.2×10^8 t oil equivalent, respectively, accounting for 34.2%, 9.8% and 33.2% of the world's total recoverable quantities of conventional oil and gas resources, respectively (Table 1.2).

1.3.3 Unconventional Oil and Gas Resources Evaluation Results

According to the resource evaluation results of RIPED in 2023, the total technical recoverable resources of unconventional oil and gas in the world are 7249.9×10^8 t oil equivalent, of which the technical recoverable resources of unconventional oil are 4244.9×10^8 t, accounting for 58.6% of the total unconventional oil and gas resources. The technically recoverable resources of unconventional natural gas are 351.6×10^{12} m^3, accounting for 41.4% of the unconventional oil and gas resources (Table 1.3).

Among the global unconventional oil, the recoverable resources of oil shale are the largest, reaching 1540.8×10^8 t, accounting for 36.3%. The recoverable heavy oil resources are 1274.8×10^8 t, accounting for 30.0%. The recoverable resources of oil sands are 638.9×10^8 t, accounting for 15.1%. The recoverable resources of shale oil are 790.4×10^8 t, accounting for 18.6%.

Shale gas constitutes the largest portion of unconventional natural gas resources in the world, and its technically recoverable resources are 235.9×10^{12} m^3, accounting for 67.1%. The recoverable resources of coalbed methane are 53×10^{12} m^3, accounting for 15.1%. The recoverable resource of tight gas is 62.7×10^{12} m^3, accounting for 17.8%.

Table 1.2 Global recoverable resources of conventional oil and gas reported by RIPED in 2023

Region	Cumulative production			Remaining recoverable reserves			Reserve growth from discovered oil and gas fields			Undiscovered recoverable resources			Total
	Oil	Condensate	Gas	Oil	Condensate	Gas	Oil	Condensate	Gas	Oil	Condensate	Gas	(10^8 t oil equivalent)
	(10^8 t)	(10^8 t)	(10^{12} m^3)	(10^8 t)	(10^8 t)	(10^{12} m^3)	(10^8 t)	(10^8 t)	(10^{12} m^3)	(10^8 t)	(10^8 t)	(10^{12} m^3)	
North America	223.48	11.08	22.86	89.48	5.26	7.18	69.77	4.07	4.23	160.08	56.01	24.58	1122.2
Central and South America	179.12	3.80	7.39	153.27	10.59	10.51	51.09	2.64	2.78	435.50	20.58	15.33	1164.3
Europe	105.72	6.85	15.61	38.39	5.27	7.07	8.79	2.75	2.43	56.45	13.68	9.56	534.3
Africa	179.58	8.72	6.93	122.20	19.96	22.85	76.06	5.20	9.87	157.07	47.44	22.91	1151.0
Middle East	489.90	14.29	6.02	911.81	105.93	112.53	167.08	34.70	22.88	296.03	47.47	39.41	3612.9
Central Asia	51.69	9.84	8.45	47.10	8.07	17.38	20.59	5.51	8.18	54.99	12.78	25.20	716.7
Russia	261.87	0.91	24.11	194.63	24.21	39.99	63.79	3.72	7.60	148.84	38.43	47.58	1755.9
Asia–Pacific	159.47	9.85	12.97	73.26	15.17	26.25	24.65	4.91	9.22	221.05	21.71	50.72	1377.6
Total	1650.83	65.34	104.34	1630.14	194.51	243.76	481.82	63.50	67.19	1530.01	258.10	235.29	11,434.90

Note 1 t crude oil = 7.14 bbl crude oil = 1170 m^3 natural gas

Table 1.3 Global recoverable resources of unconventional oil and gas reported by RIPED in 2023

Region	Technically recoverable unconventional oil resources (10^8 t)				Technically recoverable unconventional gas resources (10^{12} m^3)			Total (10^8 t oil equivalent)
	Shale oil	Heavy oil	Oil sands	Oil shale	Shale gas	CBM	Tight gas	
North America	313.59	324.70	403.39	544.53	74.32	16.98	8.60	2440.1
Central and South America	89.38	418.20	0.00	153.23	40.48	0.03	10.20	1094.2
Europe	23.63	84.24	17.93	200.09	16.69	2.01	3.73	517.6
Africa	63.35	64.77	25.00	69.68	31.66	0.59	10.60	589.0
Middle East	59.00	180.59	0.00	62.65	16.08	0.00	8.40	511.5
Central Asia	16.61	44.57	59.40	0.00	2.68	3.20	4.30	207.6
Russia	130.26	88.96	125.67	338.18	19.66	13.11	6.34	1017.3
Asia–Pacific	94.60	68.76	7.52	172.42	34.30	17.13	10.49	872.5
Total	790.47	1274.79	638.96	1540.83	235.87	53.05	62.66	7249.80

Note 1 t crude oil $= 7.14$ bbl crude oil $= 1170$ m^3 natural gas

1.4   Comparison of Authoritative Evaluation Results of Global Oil and Gas Resources

Although many international oil companies, research institutions and researchers have carried out global oil and gas resource evaluation studies, due to various constraints, it is difficult to compare the evaluation results released by different entities. Considering the timeliness of the evaluation results, the evaluation results of RIPED, USGS, and the statistical results of the IEA, which were newly released and relatively authoritative, were selected.

1.4.1   Comparison of Evaluation Results By RIPED and USGS

USGS, an international authority for predicting the quantity of global oil and gas resources, has continuously carried out oil and gas resource evaluation in major basins worldwide and issued several global oil and gas resource evaluation reports (USGS 2003, 2005, 2012).

According to the USGS evaluation, the global oil and gas zone is divided into eight regions, involving a total of about 1000 oil–gas-bearing basin provinces, of which 406 have considerable oil and gas resources. USGS introduced the concept of total petroleum system (TPS) in the evaluation report released in 2000 and subdivided the assessment unit (AU) in the system as the basic assessment unit of oil and gas resources. In terms of oil and gas geology, AU is between TPS and play. In 2000, USGS evaluated 149 TPSs and 246 assessment units in 128 major oil–gas-bearing provinces worldwide. In the new round of evaluation in 2012, the evaluation object covered 171 geological provinces and 313 assessment units. Since 2012, although the USGS has carried out oil and gas resource evaluation every year, it has not systematically evaluated and published global oil and gas resource evaluation data for a single basin.

A comparative analysis of the results of the last three global oil and gas resource evaluation released by USGS shows that the total oil and gas resources undiscovered in the world have not changed much, but the quantity of different types of oil and gas resources undiscovered has increased and decreased, and the change of condensate oil is the most obvious. Compared with the evaluation results published in 2000, the estimated global oil and condensate resources undiscovered in 2012 decreased by 13% and 19% respectively, while the natural gas resources undiscovered increased significantly by 30%, reflecting the decrease in global oil reserve resources and the increasing contribution of natural gas to global oil and gas resources. Among the predicted growth of global total oil and gas reserves in 2012, the growth of oil reserves increased slightly by 9%, while that of natural gas and condensate reserves decreased significantly by 53% and 62%, respectively, indicating that USGS revised the overestimated prediction results of reserve growth in 2000 to some extent using the new evaluation method (USGS 2012). In particular, the USGS has recognized through its years of oil and gas resource evaluation practice that due to the difficulty of predicting technological and economic changes, estimating ultimate recoverable oil and gas reserves based on current understanding is less reliable and meaningful. Considering that, in subsequent evaluations, USGS only predicted the quantity of undiscovered oil and gas resources that could be converted into reserves in existing or new oil and gas regions in the next 30 years and the quantity of reserves growth in discovered oil and gas regions (Tao 2015).

Due to the large differences in evaluation methods, time difference, evaluation scope and basin, the comparability between evaluation results of USGS and RIPED is relatively poor (Table 1.4). For example, the evaluation results of the USGS (2000) cover no basins of the Central African Rift System. When the evaluation results were updated in 2012, the five basins, including Bongor, South Chad, Muglad, Melut and Anza basin in the Central African Rift System, were collectively referred to as the total Cretaceous-Tertiary petroleum system of the Sud Basin Province and were regarded as one assessment unit, indicating that the undiscovered recoverable resources of oil, condensate and natural gas are 73.1×10^8 bbl, 3.5×10^8 bbl and 23.1×10^8 bbl oil equivalent, respectively. Therefore, no separate statistical analysis was carried out for different basins and layers in the evaluation results, and the basin and layer with the greatest potential cannot be identified. Different from the USGS, RIPED (2023) has evaluated a total of 13 plays in the above five basins in the process of evaluating the oil and gas resources of the Central African Rift System, including the Cretaceous, Paleogene and basement. The results indicate that the total oil, condensate and natural gas resources undiscovered in different basins and strata of the Central African Rift System are 85.9×10^8 bbl, 1.6×10^8 bbl and 14.5×10^8 bbl oil equivalent, respectively. During the evaluation of oil and gas resources in the Sirte Basin, the assessment units of USGS include onshore and offshore Sirte areas, and the specific resources of different strata are not discussed. RIPED has divided the Sirte Basin into 6 plays according to different formation and evaluated the quantity of resources undiscovered for each play. It can be found that there are huge differences between USGS and RIPED during the evaluation of oil and gas resources in terms of evaluation concepts and units, which have an important impact on the evaluation results.

The analysis reveals that compared with USGS (2012), the total global oil, condensate and natural gas resources estimated by RIPED (2023) were slightly larger. In addition to the quantity of condensate resources undiscovered and the growth of oil and gas reserves, RIPED (2023) released a higher evaluation result on different types of oil and gas resources. RIPED (2023) evaluated the quantity of global oil and gas resources undiscovered in a wider range of evaluation, the basic data was relatively new, and with the method using play as the basic assessment unit, the recent exploration trend and future exploration potential can be reflected comprehensively.

According to the regional division plan in the evaluation process of USGS (2012), the evaluation results of global undiscovered oil and gas resources from RIPED (2023) are divided by regions (Table 1.5). Statistical analysis shows that the evaluation report of USGS (2012) only covers 171 basin provinces and the evaluation report of RIPED (2023) covers 468 basins in the world. The evaluation methods of USGS (2012) are mainly statistical methods such as the seventh approximation

Table 1.4 Comparison of global oil and gas resources estimated by USGS in 2000 and RIPED in 2023

Evaluation index	USGS (2000)			RIPED(2023)		
Type	Oil/ $\times$ 10^8 t	Natural gas/ $\times$ 10^{12} m^3	Condensate oil/ $\times$ 10^8 t	Oil/ $\times$ 10^8 t	Natural gas/ $\times$ 10^{12} m^3	Condensate oil/ $\times$ 10^8 t
Conventional oil and gas undiscovered	1007.3	139.8	283.6	1530.0	235.3	258.1
Reserve growth (conventional)	963.2	100.2	58.8	481.8	67.2	63.5
Remaining reserves	1450.4	131.2	95.2	1630.1	243.8	194.5
Cumulative production	994.0	48.0	9.8	1650.8	104.3	65.3
Total	4414.9	419.2	447.4	5292.8	650.6	581.4

Note The quantity of undiscovered recoverable oil and gas resources from RIPED (2023), which the China data are based on Reference (Zheng 2019; Wu 2022)

model, while RIPED (2023) chooses the appropriate evaluation methods for basins with different exploration maturity. Therefore, the quantity of global oil and gas resources undiscovered estimated by RIPED is significantly greater than that of the USGS. The analysis shows that the evaluation results of RIPED (2023) are more representative and practical.

1.4.2 Comparison of Evaluation Results By RIPED and IEA

IEA, based on the investigation and summary of the oil and gas resource evaluation reports issued by different institutions worldwide, publishes data from time to time on the quantity of global oil and gas resources, mainly to build global energy models and to predict the development trend of the economy. IEA released the latest oil and gas resource evaluation results of different regions in the world in 2021 (Tables 1.6 and 1.7, up to the end of 2019), and it is marked as IEA (2019) in this paper (IEA 2021), containing the full series of conventional and unconventional data, with good reference. It is of great significance in comparison with the evaluation results of RIPED (2023).

Overall, the IEA (2019) global remaining proven recoverable reserves are similar to the 2P recoverable reserves estimated by RIPED (2023), but there are significant differences in the evaluation results within the same region mainly due to the differences in the definition of recoverable reserves by different institutions, the statistical number of discovered oil–gas fields and samples. The global recoverable conventional oil resources undiscovered from RIPED (2023) is lower than the statistical results from IEA (2019), especially in the Middle East and Africa regions. The total quantity of global unconventional oil resources from RIPED (2023) is lower than the statistical results of IEA (2019), and the difference in heavy oil resources is obvious. The total quantity of oil shale oil resources is similar, but there is a big gap between the oil shale oil resources in Europe and North America. Evaluation results of shale oil are very consistent (Table 1.6).

Similar to the evaluation results of oil resources, the global remaining recoverable natural gas reserves of IEA (2019) are similar to that of RIPED (2023), but there is a some gap in the evaluation results of the same region, mainly because different institutions' definitions of recoverable reserves and the caliber and statistical samples of discovered oil and gas fields are different. The quantity of undiscovered natural gas resources released by RIPED (2023) is significantly smaller than that of IEA (2019), especially in the Middle East and Europe. In terms of unconventional natural gas resources evaluation, the global unconventional natural gas resources in the evaluation results released by RIPED (2023) are mostly lower than that of IEA (2019), and the shale gas and coalbed methane resources are slightly lower than that of IEA (2019), while the tight gas resources are much lower. This is mainly because RIPED's tight gas resource evaluation (2023) is based on data from North America, Europe and the Asia–Pacific region, and the overall evaluation is less comprehensive (Table 1.7).

Table 1.5 Comparison of global undiscovered oil and gas resources in different regions estimated by USGS in 2012 and RIPED in 2023

Regions	USGS (2012)				RIPED (2023)			
	Number of basins	Oil/ × 10^8 t	Natural gas/ × 10^{12} m^3	Condensate oil/ 10^8 t	Number of basins	Oil/ × 10^8 t	Natural gas/ × 10^{12} m^3	Condensate oil/ × 10^8 t
Middle East–North Africa	26	321.8	38.8	114.5	35	318.5	42.5	57.1
Central Asia-Russia	29	162.4	45.6	76.7	44	203.8	72.8	51.2
North America	21	214.9	19.3	11.0	82	160.1	24.6	56.0
Central and South America	31	147.2	13.8	28.3	65	435.5	15.3	20.6
Sub-Saharan Africa–Antarctica	13	100.1	6.7	15.1	44	134.6	19.8	37.8
Asia-Pacific	39	41.7	10.7	21.5	124	210.7	46.7	18.6
South Asia	6	5.0	3.4	3.6	31	10.3	4.0	3.1
Europe	6	31.2	8.8	19.1	43	56.4	9.6	13.7
Total (global)	171	1024.3	147.1	289.8	468	1530.0	235.3	258.1

Note The quantity of undiscovered recoverable oil and gas resources from RIPED (2023), which Chian data is based on Reference (Zheng 2019; Wu 2022)

Table 1.6 Comparison of global oil resources between IEA (2019) and RIPED (2023)

Region	IEA (2019)						RIPED (2023)					
	Remaining recoverable reserves	Conventional oil undiscovered	Shale oil	Condensate oil	Heavy oil	Oil shale	Remaining recoverable reserves	Undiscovered conventional oil	Undiscovered condensate	Shale oil	Heavy oil + oil sand	Oil shale
North America	333.2	337.4	305.2	228.2	1120.0	1400.0	94.7	229.8	60.1	313.59	728.09	544.53
Central and South America	410.2	343.0	84.0	70.0	691.6	4.2	163.9	486.6	23.2	89.38	418.2	153.23
Europe	21.0	82.6	26.6	39.2	4.2	8.4	43.7	65.2	16.4	23.63	102.17	200.09
Africa	176.4	431.2	75.6	119.0	2.8	0.0	142.2	233.1	52.6	63.35	89.77	69.68
Middle East	1167.6	1264.2	40.6	225.4	19.6	42.0	1017.7	463.1	82.2	59	180.59	62.65
Central Asia - Russia	204.4	331.8	119.0	82.6	772.8	25.2	274.0	288.2	60.4	146.87	318.6	338.18
Asia-Pacific	71.4	176.4	100.8	93.8	4.2	22.4	88.4	245.7	26.6	94.6	76.28	172.42
Global	2384.2	2966.6	751.8	858.2	2615.2	1502.2	1824.6	2011.8	321.6	790.42	1913.7	1540.8

Table 1.7 Comparison of global natural gas resources between IEA (2019) and RIPED (2023) (unit: $\times 10^{12}$ m^3)

Region	IEA (2019)						RIPED (2023)					
	Remaining recoverable reserves	Undiscovered conventional	Tight gas	Shale gas	Coalbed methane	Total gas resources	Remaining recoverable reserves	Undiscovered conventional	Tight gas	Shale gas	Coalbed methane	Total gas resources
North America	16.0	50.0	10.0	81.0	7.0	149.0	7.2	28.8	8.6	74.3	17.0	135.9
Central and South America	8.0	28.0	15.0	41.0		84.0	10.5	18.1	10.2	40.5	0.0	79.3
Europe	5.0	19.0	5.0	18.0	5.0	46.0	7.1	12.0	3.7	16.7	2.0	41.5
Africa	19.0	51.0	10.0	40.0		101.0	22.8	32.8	10.6	31.7	0.6	98.5
Middle East	81.0	102.0	9.0	11.0		121.0	112.5	62.3	8.4	16.1	0.0	199.3
Central Asia - Russia	77.0	132.0	10.0	10.0	17.0	169.0	57.4	88.6	10.6	22.3	16.3	195.2
Asia–Pacific	22.0	45.0	21.0	53.0	20.0	139.0	26.3	59.9	10.5	34.3	17.1	148.1
Global	228.0	427.0	80.0	254.0	49.0	809.0	243.8	302.5	62.7	235.9	53.0	897.8

1.5 Future Development Trends and Outlook

Since the beginning of global oil and gas resource evaluation, with the continuous progress of oil and gas geological theory and exploration technology, great progress has been made in the evaluation. However, with the increasing discovery rate of oil and gas, the increasingly complex exploration objects, coupled with the impact of new energy and the new requirements of low-carbon development, the global oil and gas resource evaluation will face greater challenges in the future.

1.5.1 Oil and Gas Resources Continuously Converted into Reserves and Production in the Future

From the development history of oil and gas exploration and development, and resource evaluation, it can be found that with the continuous innovation of oil and gas geological theory and the continuous progress of exploration technology, oil and gas are constantly discovered and produced, and oil and gas resources are not exhausted. The main reasons are as follows: On the one hand, oil and gas resources in new fields, new plays or new trap types in old areas are constantly discovered, and even new mineral species are defined. On the other hand, oil and gas resources that were previously thought to be impossible to develop are being produced economically and efficiently. Driven by science and technology, the exploitation cost of oil and gas resources is bound to decrease in the future, and oil and gas resources will be an important part of the diversified resource supply to support the continuous progress of human society.

In recent years, the fields of oil and gas exploration and development have been extended and expanded from land and shallow water area to deep water regions, from medium-shallow layer to deep layer, and from conventional to unconventional oil and gas (Zou 2011, 2020). In the future, resources in the fields of ultra-deep layers, ultra-deep water, polar regions, and natural gas hydrate (Lei 2021) will be effectively utilized. It is a very important development trend to develop evaluation methods and characteristic technologies for the technical/economic recoverable oil and gas resource potential in the "two deep and one non" (deep water, deep layer and unconventional) fields, and to carry out targeted evaluation work.

1.5.2 Oil and Gas Resources Evaluation Tending to be "Holistic" with Source Rocks as the Core

According to the accumulation factors, the evaluation methods of oil and gas resources can be classified into two categories: the evaluation methods for reservoir (target layer) and the evaluation methods for source rock (Guo 2016; Wu 2005; Long

2005). Among them, the former is easy to underestimate the evaluation results, but it is easy to determine the hierarchical distribution of resources and can directly guide the exploration. The latter can be used to estimate the maximum value of resources, but it is not easy to allocate resources. Since oil and gas are generated in source rocks and stored in a basin, it is an important development trend to evaluate the oil and gas resources of the basin integrally with the basin as an integral part, the source rocks as the core and the play as the assessment unit. During the "13th Five-Year 2 Plan" period (2016–2020), the above evaluation process established during the exploration by RIPED has ensured that important oil and gas resource types in the basin are all included into the evaluation and are not double counted, but the "integrity" has not been achieved in the evaluation method and process. In the future, a set of technical processes will be further developed to accurately describe the oil and gas generation and expulsion history of source rocks and realize the overall evaluation of different types of oil and gas resources in the basin starting from the source rocks, based on the evolution history of the basin, and combined with the diagenetic history of the reservoir. Besides, combined with different evaluation methods for reservoirs, the "holistic" evaluation technology is corrected, to form an overall understanding of different types of oil and gas resources in the basin.

1.5.3 *"Three-Dimensional" Presentation of the Evaluation Results of Oil and Gas Resource*

The goal of oil and gas resource evaluation is to realize the utilization of oil and gas resources. Compared with the purpose of determining the quantity of oil and gas resources contained in a basin, it is more important to define the type and bed series distribution of oil and gas resources in the basin. In other words, it is necessary to establish a "three-dimensional" concept of resource potential and distribution. Especially for the oil and gas resources of the continuous accumulation type, it is necessary to describe the spatial distribution of resources on the basis of determining the core parameters that control the distribution of resources and identify the sweet spot and sweet spot inervals of resources, so as to lay the foundation for the next exploration, development and utilization of oil and gas resources.

1.5.4 *"Economy" and "Low Carbon" Highlighted in the Evaluation Process of Oil and Gas Resources*

With the development of new energy and the hard requirements of the goal of "Carbon Peaking and Carbon Neutrality", only the economic and clean energy with low exploration and development cost, good economy, high resource conversion rate and low carbon emissions can be effectively used in the future. Therefore, more attention should be paid to evaluating the potential and large-scale utilization of natural gas resources and developing a targeted evaluation technology suitable for natural gas. Meanwhile, the economic evaluation of different types of oil and gas resources under different oil prices should be enhanced. In general, the evaluation results of future oil and gas resources cover not only the quantity of technical recoverable resources but also the quantity of economic recoverable resources, and the environment and carbon emissions (Zou 2021).

1.5.5 Important Role of Big Data and Artificial Intelligence

In essence, the evaluation of oil and gas resources is to build a suitable prediction model through statistical analysis of the existing data, to predict the future exploration and development potential of oil and gas resources. The oil and gas industry has the typical characteristics of big data, and various oil companies have accumulated massive, big data resources in the process of exploration and development. In the future, the application of big data analysis, machine learning and other means will be strengthened in the evaluation of oil and gas resources, and artificial intelligence will be used to build more suitable prediction models, determine the scale distribution and number distribution of oil and gas reservoirs, and predict the spatial distribution of oil and gas reservoirs (Li 2018; Yang 2021), so as to make the evaluation results more accurate, reasonable, operational and instructive.

Chapter 2
Methods for Evaluating Global Oil and Gas Resources

Global oil and gas resources include both conventional and unconventional oil and gas resources. There are four parameters for evaluating conventional oil and gas resources, namely, cumulative production, remaining recoverable reserves, reserve growth from discovered oil and gas fields, and undiscovered recoverable resources. Seven main types of unconventional oil and gas resources, including shale oil/tight oil, heavy oil, oil sands, oil shale, shale gas, tight gas, and CBM have been evaluated. During this evaluation, the conventional and unconventional oil and gas resources have been considered as a whole, and a system of processes and methods have been established to evaluate all types of conventional and unconventional oil and gas resources by oil and gas accumulation play considering each basin as a whole and each petroleum system as a core.

2.1 Evaluation Ideas and Processes

2.1.1 Types and Definitions of Oil and Gas Resources

Conventional oil and gas resources refer to oil and gas resources that are stored in traps and can be extracted using existing techniques to achieve natural industrial production (free-flowing oil and gas in subsurface traps). Conventional oil and gas resources can be measured with four parameters, namely, cumulative production, remaining recoverable reserves, reserve growth from discovered oil and gas fields, and undiscovered recoverable resources. Cumulative production is the cumulative amount of all oil or gas from an oil or gas field (reservoir) that has been put into production. The data used for this evaluation is derived from S&P Global historically recorded oil and gas fields production data as of the end of 2023. Remaining recoverable reserves refer to the recoverable reserves of oil or gas that have not been recovered under current technical and economic conditions after an oil or gas field (reservoirs) has been put into production. Remaining recoverable reserves are the difference between recoverable reserves and cumulative production. According to the statistical data as of the end of 2020 obtained from the relevant data provided by S&P Global, the remaining recoverable reserves are equivalent to the 2P (proved plus probable) recoverable reserves. Reserve growth from discovered oil and gas fields refers to the additional recoverable reserves in discovered oil and gas fields due to various factors, such as progressive exploration, technological progress, and the use of different calculations methods, throughout the entire life cycle of oil and gas fields from evaluation to development since their discovery. Such reserve growth represents an important source of additional oil and gas reserves in coming decades. Considering the limitations of current technologies and understandings, the reserve growth from discovered oil and gas fields mentioned herein refers to the reserve growth potential in the next 30 years. Undiscovered recoverable oil and gas resources refer to the recoverable oil and gas resources that have not been discovered but can finally be discovered in the future with sufficient exploration works and continuous progress in exploration technologies (Table 2.1). Oil and gas resources include oil, condensate, and gas.

The unconventional oil and gas resources evaluated herein are technically recoverable resources, which refer to various types of oil and gas resources that can be recovered under current technical conditions, including shale oil/tight oil, heavy oil, oil sands, oil shale, shale gas, tight gas, and CBM. Heavy oil is a type of crude oil that is very viscous and does not flow or cannot flow easily at normal reservoir temperatures. The viscosity of heavy oil ranges between 100 and 10,000 mPa s, and its API gravity ranges from $10°$ to $20°$ (relative density within the range of 0.934–1.0). Oil sands, also known as tar sands, are sands or other rocks that contain natural bitumen and are composed of bitumen, sand, water, and clay, with viscosity higher than 10,000 mPa s and API gravity of less than $10°$ (relative density higher than 1.0 g/ cm^3). Shale oil refers to the oil that occurs in organic-rich shale formations. The siltstones, fine-grained sandstones, and carbonates in the source rocks of organic-rich shale formations do not have natural productivity or have natural production capacities lower than the lower limit of industrial oil production. Therefore, special processes and technical measures are required to achieve industrial oil production. The technically recoverable shale oil resources evaluated herein also include tight oil in adjacent reservoirs overlying and underlying source rock formations, such as tight sandstone and carbonate reservoirs. For these reservoirs, the matrix permeability at the overburden pressure is less than 0.100 mD, and individual wells normally have no natural productivity, but industrial oil production can be achieved by taking appropriate technical measures under certain economic conditions. Oil shale is a combustible organic rock with high ash content and organic matter content. The organic matter content of oil shale is high, but the organic matters contained therein are immature. Oil can be obtained from oil shale by low-temperature dry distillation. In general, the oil content of oil shale is higher than 3.5%, and their calorific value is greater than 4.18 MJ/kg. Shale gas refers to the free and adsorbed gas within organic-rich shale formations. For these shale formations, the matrix permeability at the overburden pressure is less than 0.001 mD, and individual wells normally

Table 2.1 Types and definitions of conventional and unconventional oil and gas resources

Type		Definition
Conventional oil and gas resources	Cumulative production	Cumulative amount of oil or gas produced from an oil or gas field that has been put into production. Data as of the end of 2023
	Remaining recoverable reserves	Recoverable reserves of oil or gas that have not been recovered under current technical and economic conditions after an oil or gas field (reservoir) has been put into production. The remaining reserves evaluated herein are recoverable reserves
	Reserve growth from discovered oil and gas fields	Additional recoverable reserves in discovered oil and gas fields due to various factors, such as progressive exploration, technological progress, and the use of different calculations methods, throughout the entire life cycle of oil and gas fields from evaluation to development since their discovery
	Undiscovered recoverable resources	Recoverable oil and gas resources that have not been discovered but can finally be discovered in the future with sufficient exploration works and continuous progress in exploration technologies
Unconventional oil and gas resources	Heavy oil	Oil with viscosity of 100–10,000 mPa s and API gravity of $10°$–$20°$ at normal reservoir temperatures
	Oil sands	Oil with viscosity > 10,000 mPa·s and API gravity < $10°$ at normal reservoir temperatures
	Shale oil (including tight oil)	Oil in the source rocks of organic-rich shale formations or adjacent siltstone, fine-grained sandstone, and carbonate reservoirs with matrix permeability of not more than 0.1 mD at the overburden pressure (air permeability less than 1 mD)
	Oil shale	Oil shale is a high-ash solid combustible organic rock with high but immature organic matter content. Oil can be obtained from oil shale by low-temperature dry distillation
	Shale gas	Gas within organic-rick shale formations, which is naturally occurring and continuously accumulated and stored in extensive areas of such formations. The matrix permeability of such formations at the overburden pressure is generally not more than 0.001 mD

(continued)

Table 2.1 (continued)

Type		Definition
	Tight gas	Gas in sandstone reservoirs with matrix permeability of not more than 0.1 mD at the overburden pressure or air permeability of not more than 1 mD or other tight reservoirs
	CBM	Gas generated in the coalification process and existing in coal seams or coal deposits in the adsorbed state. The methane content of CBM is generally higher than 85%

have no natural productivity, but industrial gas production can be achieved by taking appropriate technical measures under certain economic conditions. Tight gas refers to gas trapped within tight sandstones and other low-permeability reservoirs. For these reservoirs, the matrix permeability at the overburden pressure is less than 0.100 mD, and individual wells normally have no natural productivity, but industrial gas production can be achieved by taking appropriate technical measures under certain economic conditions. CBM occurs in coal seams, which is mainly adsorbed on the surfaces of coal particles and partially free in coal pores or dissolved in coalbed water. The methane content of CBM is generally higher than 85% (Table 2.1).

2.1.2 General Evaluation Idea and Process

According to the source control theory, all types of conventional and unconventional oil and gas resources are oil and gas in different phases generated by source rocks at different thermal maturity levels and distributed at different locations of oil and gas migration and conduct pathways between source rocks and traps. The core idea of the general evaluation process is to, taking all the source rocks in a basin as the starting point, evaluate and predict the spatial distribution of oil and gas resources throughout the entire life cycle of the basin. The evaluation process is shown in Fig. 2.1. The specific steps of the evaluation process are detailed below.

The first step is to determine the number and vertical distribution of all effective source rock strata in the target basin while considering the basin as a whole.

The second step is to determine the degrees of development of oil and gas carrier systems connected with various source rock strata. The oil and gas trapped in a source rock stratum due to hindered oil and gas expulsion resulting from the lack of communication between faults and source rocks and the tight top and bottom seals of the source rock stratum are oil and gas inside the source. The oil and gas expelled from a source rock stratum under dynamic forces predominated by pressurization during oil and gas generation due to the development of fault in source rocks or non-tight top and bottom seals of the source rock stratum are oil and gas outside the source.

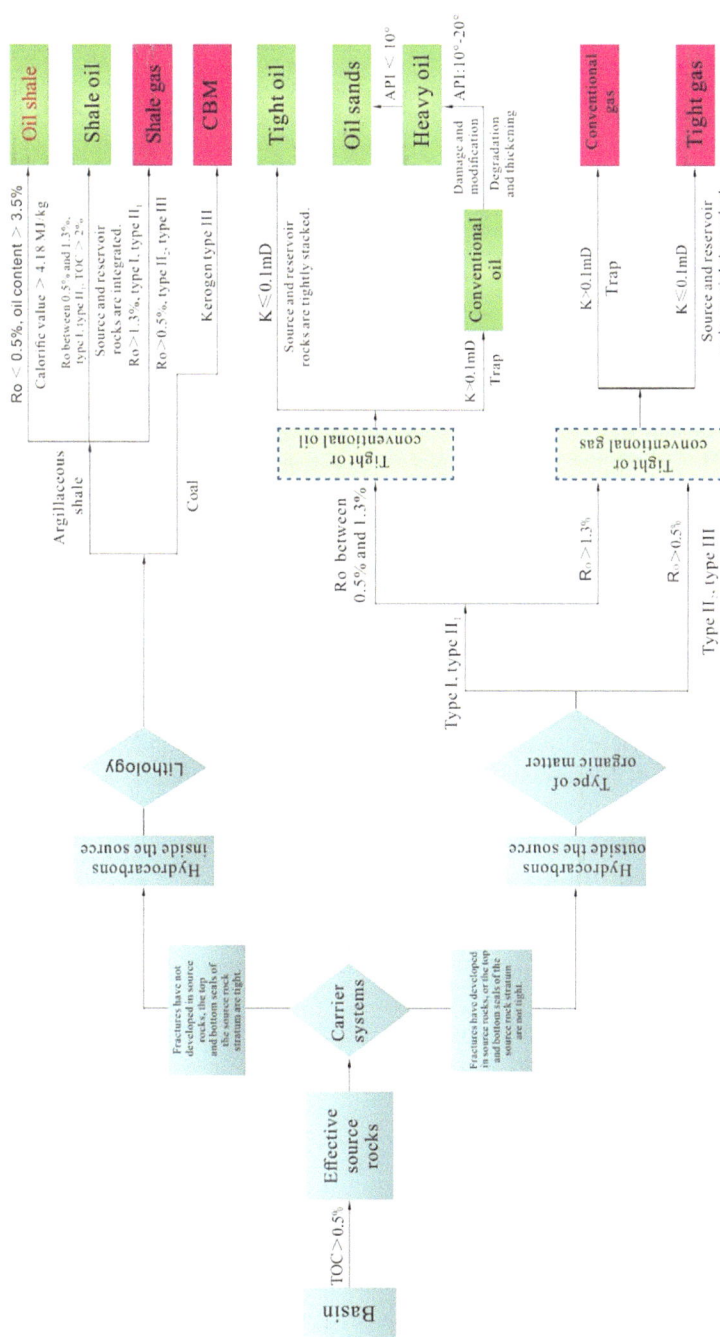

Fig. 2.1 Screening criteria and process for evaluating all types of conventional and unconventional oil and gas resources while considering each basin as a whole

The third step is to classify source rock strata into argillaceous shales (including argillaceous limestones) and coal-bearing strata depending on lithology and further divide argillaceous shales into oil-prone types I and II_1 and gas-prone types II_2 and III depending on the type of organic matter.

The fourth step is to classify oil and gas in argillaceous shales into oil shale, shale oil, and shale gas according to the value of the thermal maturity indicator R_o. For oil shale, the value of R_o is smaller than 0.5%, the oil content is higher than 3.5%, and the calorific value is greater than 4.18 MJ/kg. For shale oil, R_o ranges between 0.5% and 1.3%, the type of organic matter is type I or type II_1, and the permeability is lower than 0.1 mD. Shale gas resources can exist in formations with type I or type II_1 organic matter and R_o greater than 1.3% and/or formations with type II_2 or type III organic matter and R_o greater than 0.5%. CBM can occur in coal seams with R_o greater 0.5%. Depending on thermal maturity, the oil and gas inside type I and II_1 source rock described in the third step can be classified into tight or conventional oil and gas. The oil and gas inside source rock with R_o ranging between 0.5 and 1.3% are to be considered tight or conventional oil, and the oil and gas inside source rock with R_o greater than 1.3% are to be considered tight or conventional gas. The oil and gas inside type II_2 and type III source rock with Ro greater than 0.5% are to be considered tight or conventional gas.

The fifth step is to further classify tight or conventional oil and gas depending on reservoir permeability. Oil and gas are to be considered tight oil if reservoir permeability is less than 0.1 mD and conventional oil when reservoir permeability is higher than 0.1 mD. After conventional oil is biodegraded, thickened and/or modified in any other form, the generated oil with API gravity ranging between 10° and 20° is to be considered as heavy oil. After heavy oil is further destroyed and biodegraded, the generated oil with API gravity less than 10° is to be considered an oil resource from oil sands. The gas in reservoirs is to be considered as tight gas when reservoir permeability is lower than 0.1 mD and conventional gas when reservoir permeability is higher than 0.1 mD.

This evaluation and screening process can avoid any omission or repeated calculation of all important types of oil and gas resources in a basin and allows for an overall evaluation and understanding of the resource potential of the basin. It is to be noted that non-thermogenic gases (biogenic gases) and gas hydrates in shallow layers or frozen soils are not included in this general evaluation plan due to their special origins and production methods.

2.2 Evaluation Methods

The cumulative production and remaining recoverable reserves evaluated herein have been determined through a statistical analysis of S&P Global's data as of the end of 2020. Undiscovered recoverable resources, reserve growth from discovered oil and gas fields, and technically recoverable unconventional resources have been determined through an independent evaluation using applicable evaluation methods.

2.2.1 Methods for Evaluating Undiscovered Conventional Oil and Gas Resources

Undiscovered conventional oil and gas resources have been evaluated by oil and gas accumulation play (which is considered the basic geological unit to be evaluated) using different methods depending on the extent of exploration and data availability. The parameters of all evaluation methods were selected based on comprehensive geological evaluations (Fig. 2.2). Geological units explored to a great extent have been evaluated using the discovery process method. Petroleum asset areas and areas with more available data have been evaluated using the trap summation method. Geological units explored to a moderate extent have been evaluated using the subjective probability method based on geological analysis. Geological units with no oil or gas discovery or very limited available data have been evaluated using the multi-parameter analog method. Finally, the evaluation results obtained from various evaluation methods with respect to different geological units have been summed and summarized using the Monte Carlo simulation method. The evaluation results are expressed in the form of probability, and their confidence levels are expressed as 95%, 50%, 5%, and the mean (Mean), respectively.

During the "13th Five-Year Plan" period, the RIPED built a platform for managing the basic data of potential traps worldwide based on the ArcGIS Platform using the commercial data purchased from S&P global, data from evaluations of more than 100 new projects initiated each year, data disclosed by multi-user geophysical services companies, information disclosed by oil companies concerning subsurface traps in their operation areas, and data from other sources. On the ArcGIS platform, a global potential trap basic data management platform has been established that includes the area, depth, oil/gas reservoir potential trap resources and other basic data of 32,000 potential traps worldwide, especially for the coverage of multi-user seismic data. For offshore areas involving multiple users and high degrees of coverage by seismic data, the resources in more basins have been evaluated and calculated using the trap summation method, and the evaluation results can be used directly for the advance selection of oil and gas-bearing plays and the evaluation of new projects.

2.2.2 Methods for Evaluating Reserves Growth from Discovered Oil and Gas Fields

Reserve growth from discovered oil and gas fields refers to the additional recoverable reserves in discovered oil and gas fields due to various factors, such as progressive exploration, technological progress, and the use of different calculations methods, throughout their entire life from evaluation to development since their discovery. 70% of the world's yearly additional recoverable reserves come from discovered oil and gas fields (S&P global). Because continuous reserves data over a period spanning several years cannot be obtained from a single oil or gas field, it is difficult to

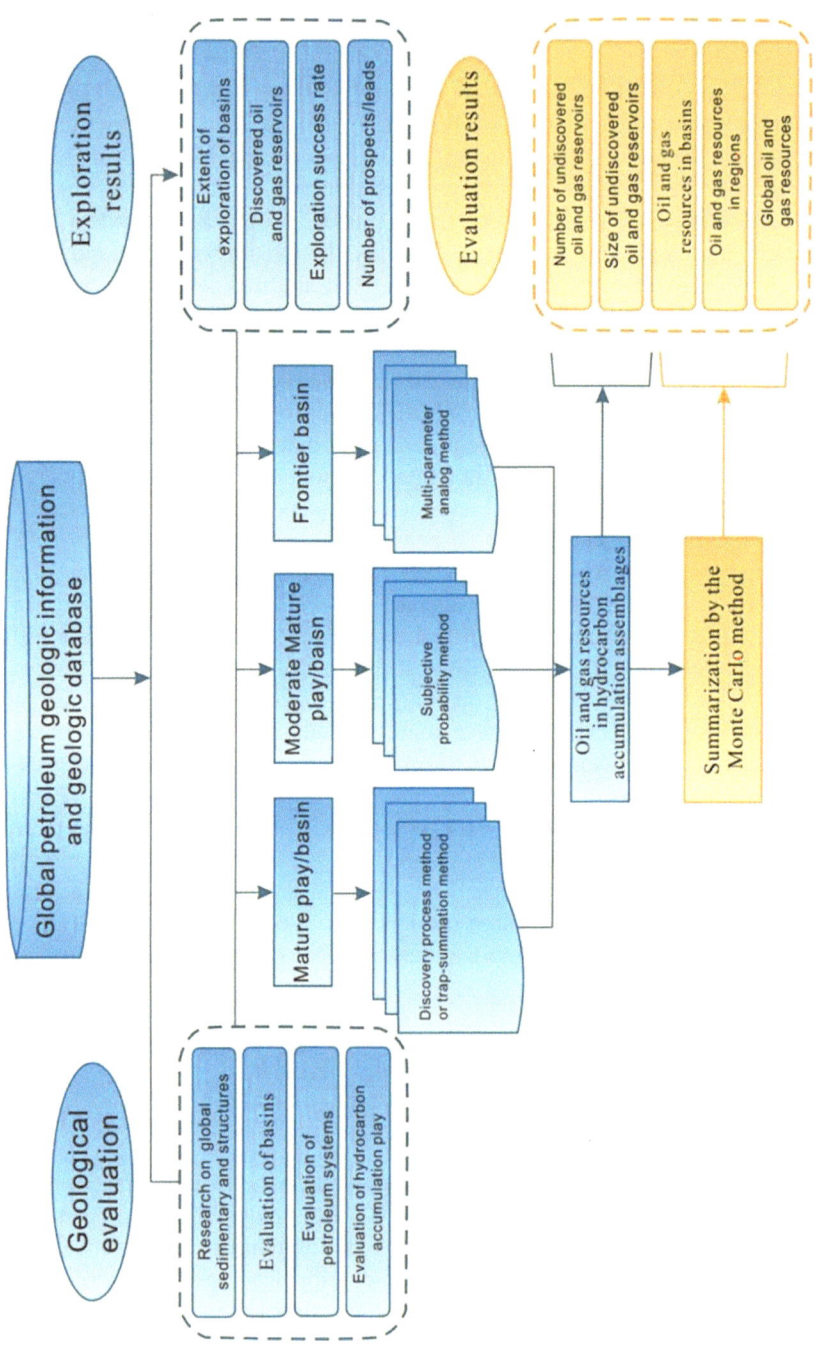

Fig. 2.2 Flowchart for the evaluation of undiscovered conventional oil and gas resources worldwide

establish a continuous reserve growth function for a certain oil or gas field. Therefore, the stepwise cumulative multiplication method has been used to find the coefficient of cumulative reserve growth from different oil and gas fields over a period of 30 consecutive years, and models for reserve growth from discovered oil and gas fields have been built. For different regions with oil and gas resources, the potentials for reserve growth from discovered oil and gas fields have been predicted using the reserve growth models built for such regions. Such reserve growth function, which has been established by taking the reserve growth in each region as a sample, is more targeted and accurate and can deliver more reasonable evaluation results. The USGS has predicted the potential for global reserve growth using only one type of model applicable to North America, which has relatively low applicability and confidence levels (Fig. 2.3). The potential for reserve growth from discovered oil and gas fields during the "13th Five-Year Plan" period has been evaluated using the latest data of recoverable reserves in discovered oil and gas fields and the reserve growth curves drawn during the "12th Five-Year Plan" period.

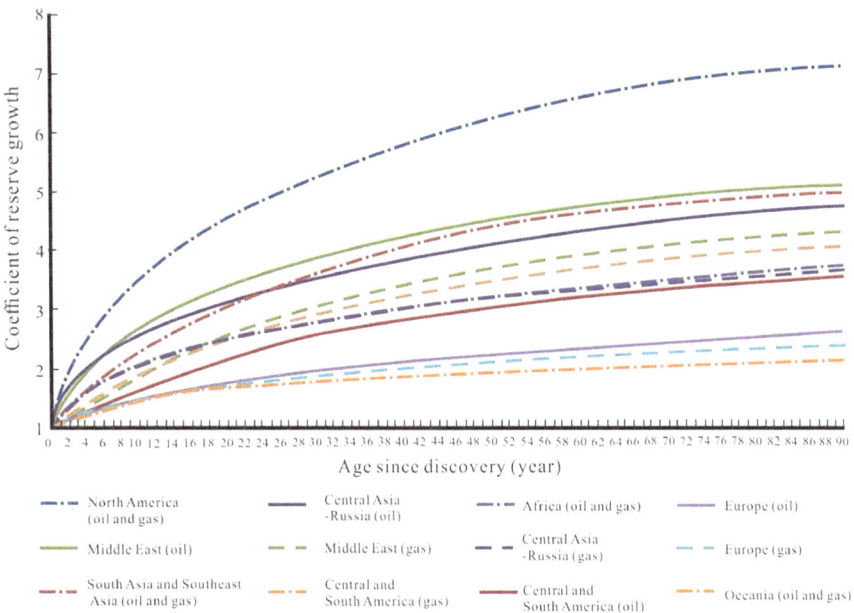

Fig. 2.3 Curves of reserve growth from discovered oil and gas fields applicable to different regions

2.2.3 Methods for Evaluating Unconventional Oil and Gas Resources

For the evaluation of global unconventional oil and gas resources, basins have been classified into three categories, namely, basins for general evaluation, basins for detailed evaluation, and key basins to be evaluated, depending on the level of detail of basin data, resource type, the extent of exploration and development, evaluation needs, and the applicability of evaluation methods. Basins have been evaluated using the parametric probability method, the GIS-based spatial interpolation method, the origin constrained volumetric method, and the hyperbolic and exponential decline method. The basins for general evaluation are mostly basins characterized by a small extent of exploration and development and a lack of basic data, parameters and maps. Such basins have been evaluated using the parametric probability method. The basins for detailed evaluation are those characterized by ongoing exploration and develop-ment activities, abundant basic geological data, and insufficient production data of production wells. The key basins to be evaluated are those characterized by ongoing exploration and commercial development activities, abundant basic geological data, large scale of resources, and sufficient detailed production data of production wells. Five types of oil and gas resources, namely, heavy oil, oil sands, oil shale, tight gas, and CBM, have been evaluated using the GIS-based spatial interpolation method. The key basins to be evaluated need to be evaluated comprehensively from the perspec-tives of resource abundance, recoverability, and cost-effectiveness for the selection of favorable plays. The oil and gas accumulation plays controlled by source rocks in basins for detailed evaluation and key basins to be evaluated, such as shale oil and shale gas, have been evaluated using the origin constrained volumetric method. Key basins characterized by a great extent of exploration and development and suffi-cient detailed production data of production wells have been evaluated using the hyperbolic and exponential decline method, the effective evaluation areas have been determined using basic geological parameters and maps, and the final recoverable reserves in favorable plays have been calculated. The detailed evaluation methods and evaluation process are shown in Fig. 2.4.

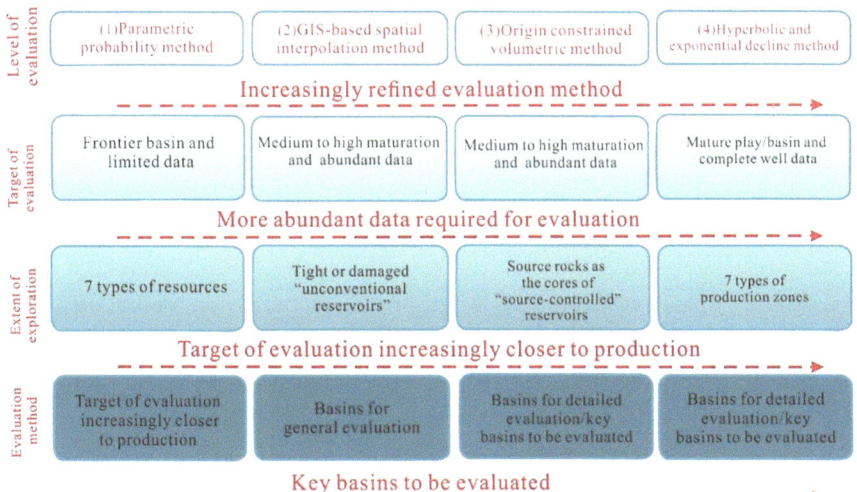

Fig. 2.4 Flowchart for the evaluation of global unconventional oil and gas resources

Chapter 3
Characteristics and Distributions of Global Oil and Gas Resources

The total recoverable conventional oil and gas resources worldwide are 11,434.9 $\times 10^8$ tons of oil equivalent, which are mainly concentrated in the Middle East, Russia, Central and South America, and Asia–Pacific. The total technically recoverable unconventional oil and gas resources worldwide are 7249.9 $\times 10^8$ tons of oil equivalent, which are mainly concentrated in North America, Central and South America, and Russia.

3.1 Conventional Oil and Gas Resources

The global recoverable conventional oil resources are 5292.8 $\times 10^8$ tons, condensate are 581.4 $\times 10^8$ tons and 650.6 $\times 10^{12}$ m^3 of gas. The cumulative oil and gas production are 2608.0 $\times 10^8$ tons of oil equivalent, and the recovery rate is 22.8%. The remaining recoverable oil and gas reserves are 3908.1 $\times 10^8$ tons of oil equivalent, accounting for 34.2% of the total recoverable resources. The reserve growth from discovered oil and gas fields are 1119.6 $\times 10^8$ tons of oil equivalent, accounting for 9.8% of the total recoverable resources. The undiscovered recoverable oil and gas resources are 3799.2 $\times 10^8$ tons of oil equivalent, accounting for 33.2% of the total recoverable resources.

3.1.1 Characteristics of Distributions of Remaining Recoverable Reserves

The remaining recoverable reserves of oil and gas worldwide are 3908.1 $\times 10^8$ tons of oil equivalent, including 1630.1 $\times 10^8$ tons of oil, accounting for 41.7%, 194.5 $\times 10^8$ tons of condensate, accounting for 5.0%, and 243.8 $\times 10^{12}$ m^3 of gas, accounting

L. Dou et al., *Global Oil and Gas Resources: Potential and Distribution*,
https://doi.org/10.1007/978-981-97-4756-6_3

for 53.3%. The remaining recoverable reserves distributed in the Middle East account for 50.7% of the total remaining recoverable reserves, and the remaining recoverable reserves in Russia and Africa account for 14.3% and 8.6% of the total remaining recoverable reserves, respectively.

3.1.1.1 Distribution of Remaining Recoverable Reserves in Major Countries (Regions)

The remaining recoverable reserves of oil and gas worldwide are distributed in 82 countries. The remaining recoverable reserves distributed in Russia, Qatar, Saudi Arabia, and Iran account for 14.3%, 12.7%, 12.5% and 11.3%, respectively. The remaining recoverable oil reserves are mainly concentrated in Saudi Arabia and Iraq, and the remaining recoverable oil reserves in these two countries account for 23.5 and 12.4% of the world's total remaining recoverable oil reserves. The remaining recoverable gas reserves are mainly concentrated in Qatar and Russia, the remaining recoverable gas reserves in these two countries account for 20.7% and 16.4% of the world's total remaining recoverable gas reserves, respectively (Figs. 3.1 and 3.2).

The remaining recoverable reserves of oil and gas in Russia are 560.6×10^8 tons of oil equivalent, of which oil accounts for 34.7%, condensate accounts for 4.3%, and gas accounts for 61.0%. The remaining recoverable reserves of oil and gas in Qatar rank second and are 494.8×10^8 tons of oil equivalent, of which oil accounts for 1.1%, condensate accounts for 11.8%, and gas accounts for 87.1%. The remaining recoverable reserves of oil and gas in Saudi Arabia are 488.7×10^8 tons of oil equivalent, of which the remaining recoverable reserves of oil account for 78.3%. The remaining recoverable reserves of oil and gas in Iran are 443.1×10^8 tons of

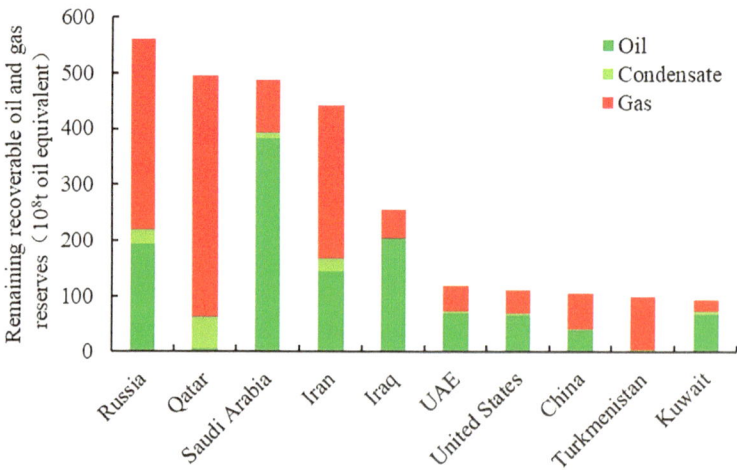

Fig. 3.1 Histogram of the remaining recoverable reserves in major countries (regions) worldwide

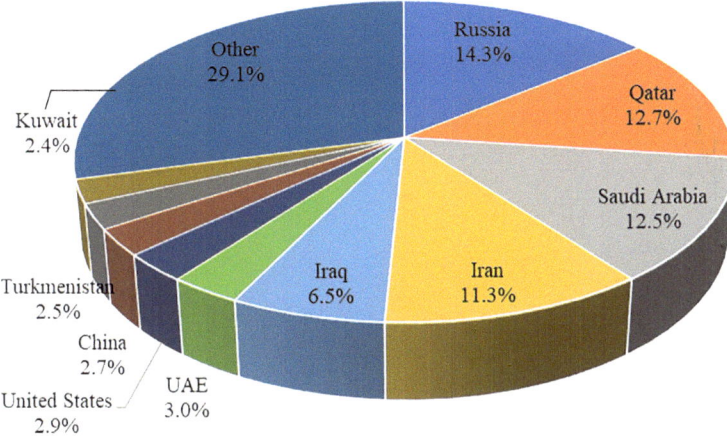

Fig. 3.2 Pie chart of the remaining recoverable reserves in major countries (regions) worldwide

oil equivalent, of which the remaining recoverable reserves of oil account for 32.5% and the remaining recoverable reserves of gas account for only 61.9%.

3.1.1.2 Distribution of Remaining Recoverable Reserves in Major Basins

The remaining recoverable reserves of oil and gas worldwide are mainly concentrated in 46 basins (more than 10×10^8 tons of oil equivalent). The remaining recoverable reserves of oil and gas in the Arabian Basin, West Siberian Basin and Zagros Basin account for 58.5% of the world's total remaining recoverable reserves (Figs. 3.3 and 3.4).

The remaining recoverable reserves of oil and gas in the Arabian Basin are 1615.9 $\times 10^8$ tons of oil equivalent, of which oil accounts for 46.0%, condensate accounts for 5.6%, and gas accounts for 48.4%. The remaining recoverable reserves of oil and gas in the West Siberian Basin rank second and are 364.0×10^8 tons of oil equivalent, of which oil accounts for 32.7%, and the remaining recoverable reserves of condensate and gas account for only 3.7% and 63.6%, respectively. The remaining recoverable reserves of oil and gas in the Zagros Basin are 306.7×10^8 tons of oil equivalent, of which oil accounts for 50.1%, and the remaining recoverable reserves of condensate and gas account for 3.8% and 46.1%, respectively.

3.1.1.3 Distribution of Remaining Recoverable Reserves in Onshore and Offshore Areas

The remaining recoverable reserves of onshore oil and gas worldwide are 2127.6 $\times 10^8$ tons of oil equivalent, of which oil accounts for 56.5%, condensate accounts for

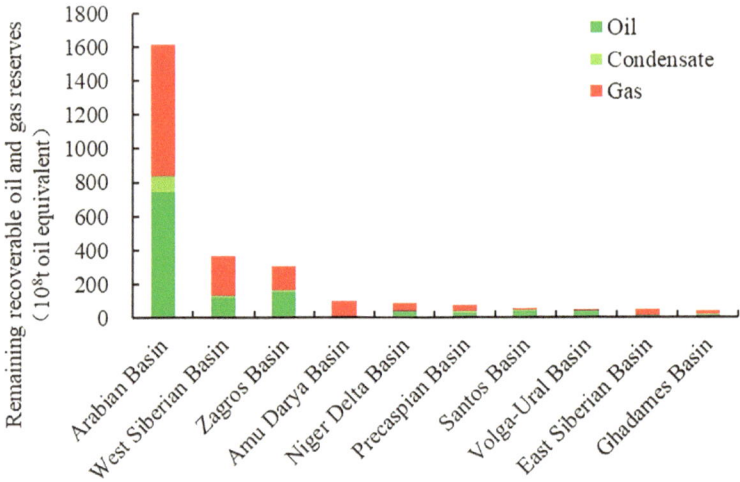

Fig. 3.3 Histogram of the remaining recoverable reserves in major basins worldwide

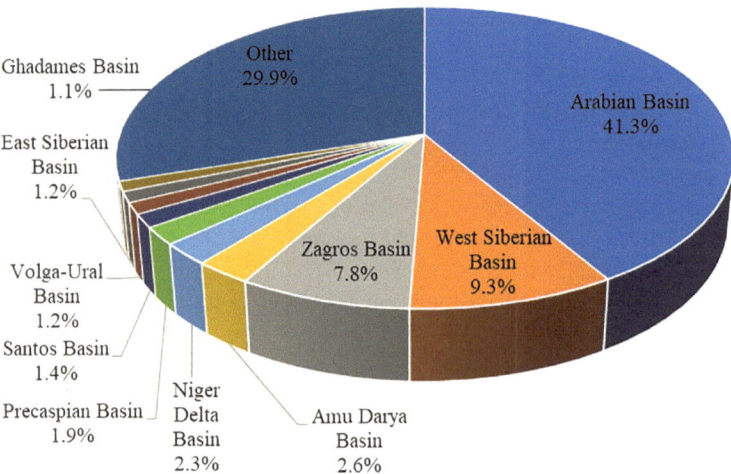

Fig. 3.4 Pie chart of the remaining recoverable reserves in major basins worldwide

3.6%, and gas accounts for 39.9%. The remaining recoverable reserves of offshore oil and gas worldwide are 1780.4×10^8 tons of oil equivalent, of which oil accounts for 31.5%, condensate accounts for 4.9%, and gas accounts for 63.6%. The remaining recoverable reserves of onshore oil and gas are larger than those of offshore oil and gas. The remaining recoverable reserves of onshore and offshore oil and gas resources account for 54.4% and 45.6% of the total remaining recoverable reserves, respectively (Fig. 3.5).

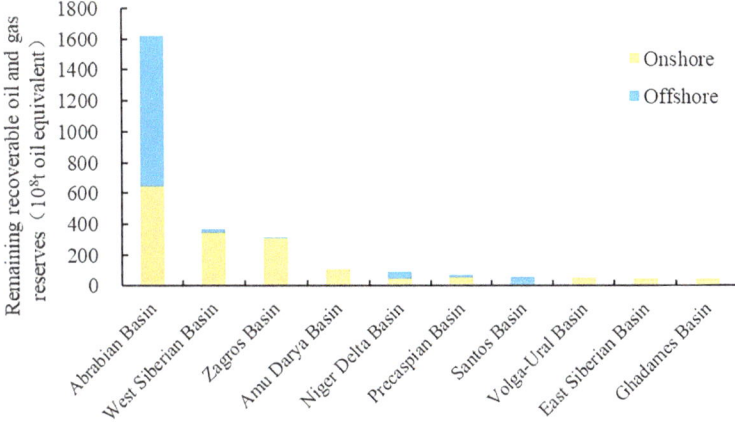

Fig. 3.5 Histogram of the remaining recoverable reserves of onshore and offshore oil and gas resources in major basins worldwide

3.1.1.4   Distribution of Remaining Recoverable Reserves in Different Types of Basins

According to the Wilson Cycle, basins worldwide can be classified into six types, including rift basins, passive margin basins, cratonic basins, back-arc basins, fore-arc basins, and foreland basins. The remaining recoverable reserves of oil and gas worldwide are mainly distributed in three types of basins, namely, foreland basins, passive margin basins, and rift basins. The remaining recoverable reserves in these three types of basins account for 66.1%, 14.9% and 12.7% of the total remaining recoverable reserves, respectively. The sum of remaining recoverable reserves in cratonic, back-arc and fore-arc basins accounts for only 6.3% of the total remaining recoverable reserves (Fig. 3.6).

The remaining recoverable reserves in foreland basins are mainly distributed in the Arabian Basin, Zagros Basin, Volga-Ural Basin, Eastern Venezuela Basin, and Mara-caibo Basin. The remaining recoverable reserves in rift basins are mainly distributed in the West Siberian Basin, Amu Darya Basin, North Sea Basin, Sirte Basin, and Malay Basin. The remaining recoverable reserves in passive margin basins are mainly distributed in the Niger Delta Basin, Santos Basin, Rovuma Basin, Lower Congo Basin, and East Barents Sea Basin.

3.1.1.5   Distribution of Remaining Recoverable Reserves in Various Formations

The remaining recoverable reserves of oil and gas worldwide are mainly distributed in the Cretaceous, Jurassic, Permian, Paleogene, and Neogene formations. The remaining recoverable reserves of oil and gas in these formations account for 27%,

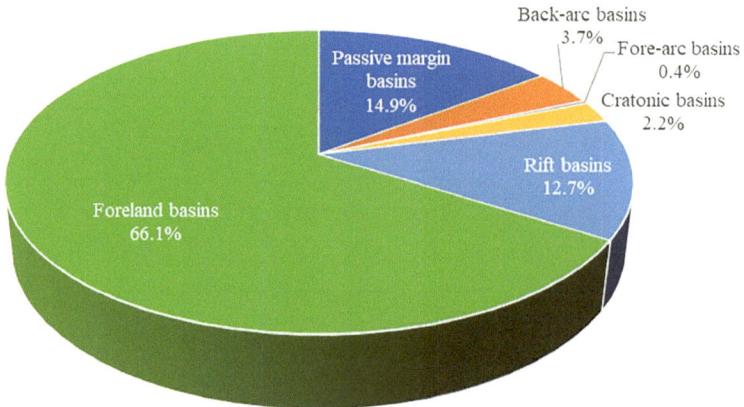

Fig. 3.6 Pie chart of remaining recoverable reserves in different types of basins worldwide

18.9%, 17.5%, 14.4% and 11.7% of the total remaining recoverable reserves, respectively. The remaining recoverable reserves of oil and gas in other formations account for only 10.5% of the total remaining recoverable reserves (Fig. 3.7).

The remaining recoverable reserves in the Cretaceous formations are mainly distributed in the Arabian Basin, West Siberian Basin, Zagros Basin, Santos Basin, and Sirte Basin. The remaining recoverable reserves in the Jurassic formations are mainly distributed in the Arabian Basin, Amu Darya Basin, West Siberian Basin,

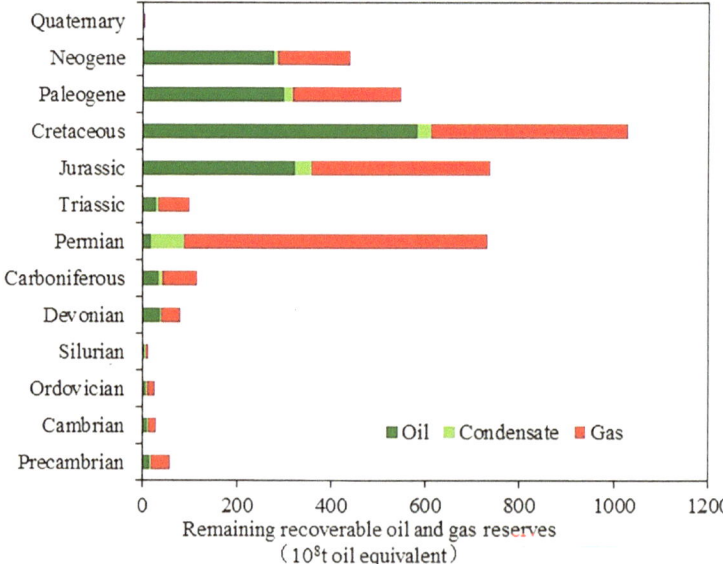

Fig. 3.7 Distribution of the remaining recoverable reserves of oil and gas in geological age worldwide

East Barents Sea Basin, and North Sea Basin. The remaining recoverable reserves in the Permian formations are mainly distributed in the Arabian Basin, Zagros Basin, Alaska North Slope, and Volga-Ural Basin. The remaining recoverable reserves in the Precambrian formations are mainly distributed in the Precambrian reservoirs of the East Siberian Basin, the Oman Basin, and some other basins.

3.1.2 Trend of Reserve Growth from Discovered Oil and Gas Fields

The reserve growth in discovered oil and gas fields worldwide for the next 30 years is 1119.6×10^8 tons of oil equivalent, of which oil accounts for 43.0%, condensate accounts for 5.7%, and gas accounts for 51.3%. The reserve growth from discovered oil and gas fields in the Middle East is the largest accounting for 35.0% of the total global reserve growth. The reserve growth of discovered oil and gas fields in Africa and Russia account for 14.8% and 11.8% of the total global reserve growth, respectively. The Asia–Pacific, North America, Central Asia, and Central and South America have similar potentials for reserve growth, while Europe has the lowest potential for reserve growth.

3.1.2.1 Distribution of Reserve Growth in Discovered Oil and Gas Fields in Major Countries (Regions)

The discovered oil and gas fields in Russia have the greatest potential for reserve growth, which are 132.5×10^8 tons of oil equivalent, accounting for 11.8% of the total global reserve growth. The reserve growth of oil and gas in these oil and gas fields are similar and account for 48.1% and 49.0% of the total global reserve growth, respectively. The reserve growth from discovered oil and gas fields in Saudi Arabia accounts for 9.2% of the total global reserve growth. The discovered oil and gas fields in Qatar and Iran have similar potentials for reserve growth, which account for 6.9 and 6.6% of the total global reserve growth. The reserve growth in discovered oil and gas fields in the United States of America (USA) and Turkmenistan account for 5.7% and 5.3% of the total global reserve growth, respectively (Figs. 3.8 and 3.9).

3.1.2.2 Distribution of Reserve Growth in Discovered Oil and Gas Fields in Major Basins

The global reserve growth of oil and gas mainly comes from discovered oil and gas fields in 32 basins, including the Arabian Basin, Zagros Basin, West Siberian Basin, Amu Darya Basin, Rovuma Basin, Nile Delta Basin, Gulf of Mexico Deepwater Basin, Sureste Basin, East Barents Sea Basin, and Tanzania Basin. The Arabian

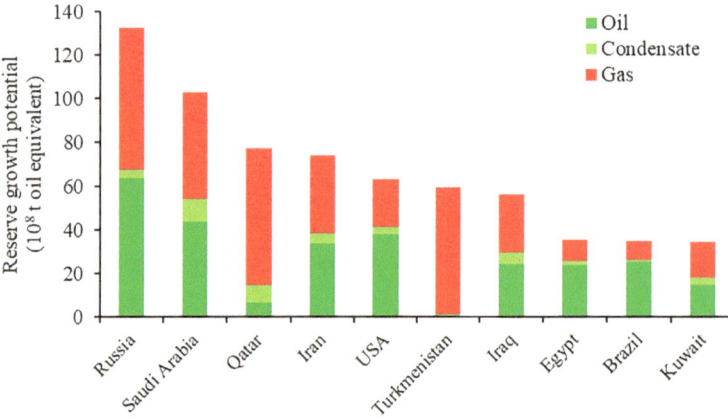

Fig. 3.8 Histogram of future reserve growth from discovered oil and gas fields in major countries (regions) worldwide

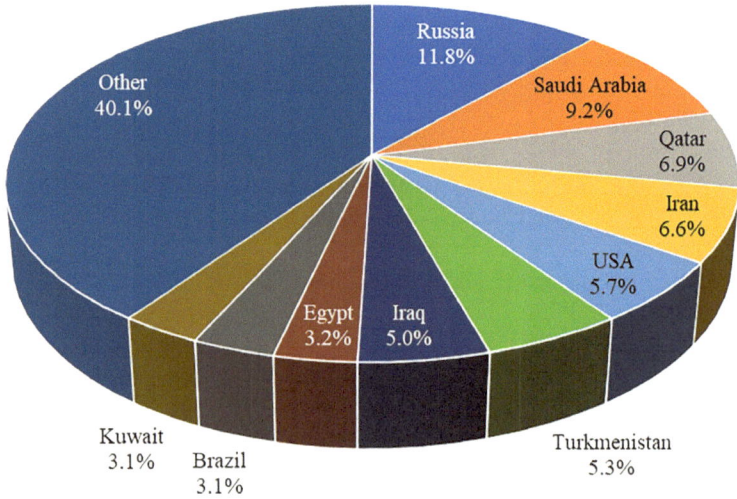

Fig. 3.9 Pie chart of future reserve growth from discovered oil and gas fields in major countries (regions) worldwide

Basin, Zagros Basin, and West Siberian Basin rank among the top three in terms of reserve growth, and the total reserve growth from discovered oil and gas fields in these three basins accounts for 41.5% of the total global reserve growth (Figs. 3.10 and 3.11).

The reserve growth from discovered oil and gas fields in the Arabian Basin are 286.4×10^8 tons of oil equivalent, of which oil accounts for 42.6%, condensate accounts for 9.9%, and gas accounts for 47.5%. The reserve growth from discovered oil and gas fields in the Zagros Basin are 92.7×10^8 tons of oil equivalent, of which

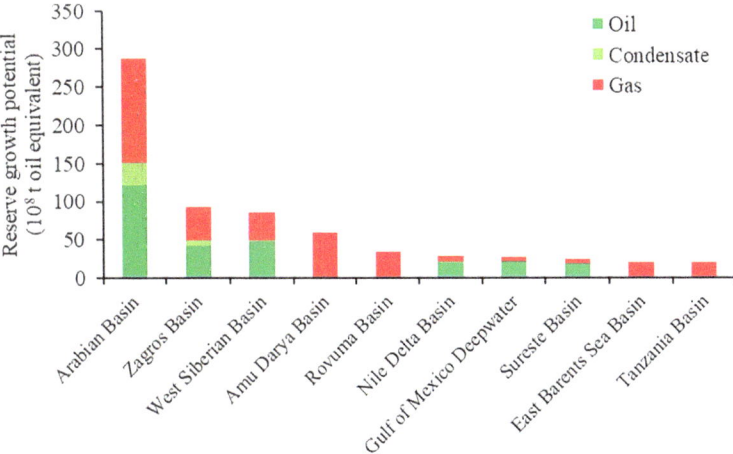

Fig. 3.10 Histogram of future reserve growth from discovered oil and gas fields in major basins worldwide

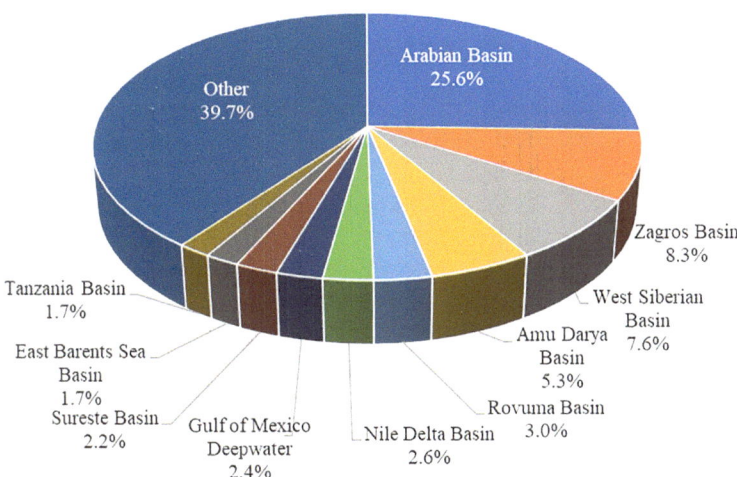

Fig. 3.11 Pie chart of future reserve growth from discovered oil and gas fields in major basins worldwide

oil accounts for 45.8%, condensate accounts for 6.5%, and gas accounts for 47.7%. The reserve growth from discovered oil and gas fields in the West Siberian Basin are 85.2×10^8 tons of oil equivalent, of which oil accounts for 55.8%, condensate accounts for 2.0%, and gas accounts for 42.2%.

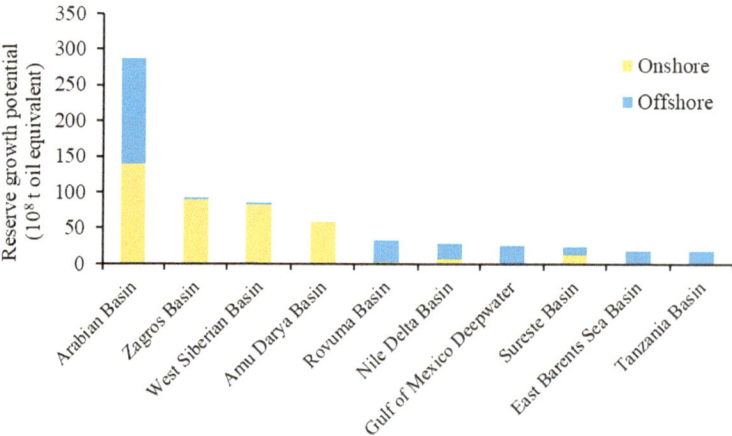

Fig. 3.12 Histogram of future reserve growth from discovered onshore and offshore oil and gas fields of major basins worldwide

3.1.2.3 Distribution of Reserve Growth in Discovered Onshore and Offshore Oil and Gas Fields

The global reserve growth from discovered onshore oil and gas fields is 631.2×10^8 tons of oil equivalent, of which oil accounts for 59.4%, condensate accounts for 3.1%, and gas accounts for 37.5%. The global reserve growth from discovered offshore oil and gas fields is 488.4×10^8 tons of oil equivalent, of which oil accounts for 39.4%, condensate accounts for 4.3%, and gas accounts for 56.3%. The reserve growth from discovered onshore oil and gas fields are greater than that from discovered offshore oil and gas fields. The reserve growth in discovered onshore and offshore oil and gas fields accounts for 56.4% and 43.6% of the total global reserve growth, respectively (Fig. 3.12).

3.1.2.4 Distribution of Reserve Growth in Discovered Oil and Gas Fields in Different Types of Basins

The reserve growth in discovered oil and gas fields worldwide are mainly distributed in three types of basins, namely, foreland basins, passive margin basins, and rift basins. The reserve growth in discovered oil and gas fields in these three types of basins account for 48.8%, 24.7% and 20.7% of the total global reserve growth, respectively. The sum of reserve growth in cratonic, back-arc and fore-arc basins accounts for only 5.8% of the total global reserve growth (Fig. 3.13).

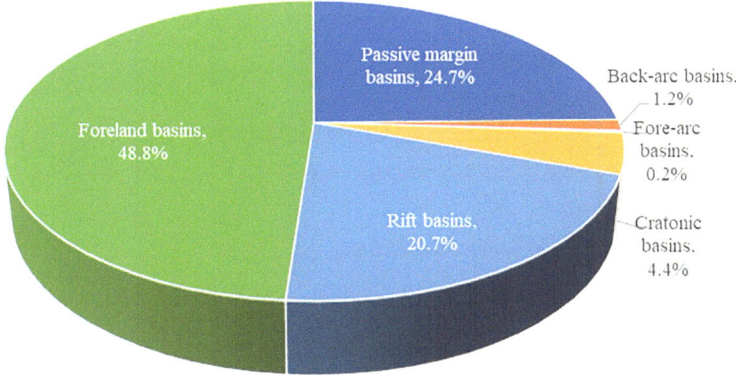

Fig. 3.13 Pie chart of global reserve growth in discovered oil and gas fields in different types of basins

The reserve growth in discovered oil and gas fields in foreland basins are mainly distributed in the Arabian Basin, Zagros Basin, Sureste Basin, Eastern Venezuela Basin, and South Caspian Basin. The reserve growth in discovered oil and gas fields in passive margin basins are mainly distributed in the Rovuma Basin, Nile Delta Basin, Gulf of Mexico Deepwater Basin, East Barents Sea Basin, and Tanzanian Basin. The reserve growth in discovered oil and gas fields in rift basins are mainly distributed in the West Siberian Basin, Amu Darya Basin, Sirte Basin, Northeast Germany-Polish Basin, and Kutei Basin.

3.1.2.5 Distribution of Reserve Growth in Discovered Oil and Gas Fields in Geological Age

The reserve growth in discovered oil and gas fields worldwide are distributed in the Cretaceous, Jurassic, Paleogene, Permian, and Neogene formations (these formations are arranged in descending order depending on the volumes of reserve growth contained herein). The reserve growth in discovered oil and gas fields in these formations account for 28.5%, 19.7%, 14.9%, 13.7% and 11.0% of the total global reserve growth, respectively. The reserve growth in discovered oil and gas fields in other formations are relatively small, accounting for only 12.2% of the total global reserve growth (Fig. 3.14).

The reserve growth in discovered oil and gas fields in the Cretaceous formations worldwide are mainly distributed in the Arabian Basin, West Siberian Basin, Zagros Basin, Santos Basin, and Sureste Basin. The reserve growth in discovered oil and gas fields in the Jurassic formations worldwide are mainly distributed in the Arabian Basin, West Siberian Basin, Amu Darya Basin, North Sea Basin, and Sureste Basin. The reserve growth in discovered oil and gas fields in the Paleogene formations worldwide are mainly distributed in the Zagros Basin, Niger Delta Basin, Eastern Venezuela Basin, Maracaibo Basin, and Gulf Coast Basin of the USA.

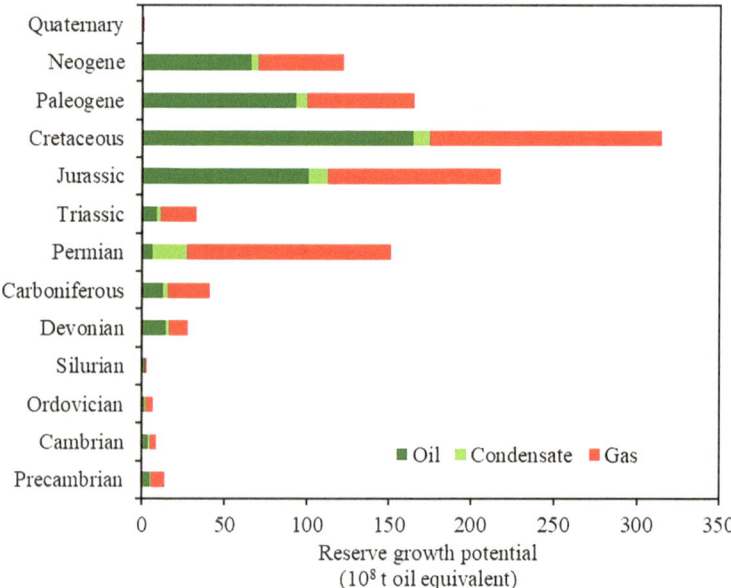

Fig. 3.14 Histogram of reserve growth in discovered oil and gas fields in geological age worldwide

3.1.3 Characteristics of Distributions of Undiscovered Oil and Gas Resources

The global undiscovered oil and gas resources are 3799.2×10^8 tons of oil equivalent, of which oil accounts for 40.3%, condensate accounts for 6.8%, and gas accounts for 52.9%. These oil and gas resources are mainly concentrated in the Middle East. The undiscovered oil and gas resources in the Middle East accounts for 17.9% of the total undiscovered oil and gas resources worldwide. The undiscovered oil and gas resources distributed in Asia–Pacific and Russia account for 17.8% and 15.6% of the total undiscovered oil and gas resources worldwide, respectively. There are also undiscovered oil and gas resources distributed in Central and South America, North America and Africa. The proportions of undiscovered oil and gas resources in Central Asia and Europe are relatively small.

3.1.3.1 Distribution of Undiscovered Recoverable Resources in Major Countries (Regions)

The undiscovered recoverable resources in Russia have the greatest potential and reach 594.0×10^8 tons of oil equivalent, of which undiscovered recoverable gas accounts for 68.5%, and undiscovered recoverable oil and condensate account for 25.1% and 6.4%, respectively. The undiscovered recoverable resources in China rank second and are 491.4×10^8 tons of oil equivalent (Zheng 2019; Wu 2022), of which undiscovered recoverable oil accounts for 32.6%, and undiscovered recoverable gas account for 67.4%. The undiscovered recoverable resources in Brazil are 346.8×10^8 tons of oil equivalent, of which undiscovered recoverable oil accounts for 82%, undiscovered recoverable condensate accounts for 1.3%, and undiscovered recoverable gas accounts for 16.7% (Figs. 3.15 and 3.16).

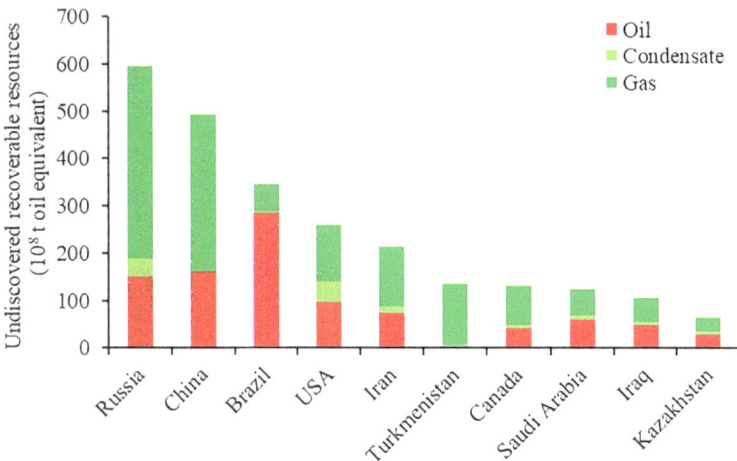

Fig. 3.15   Histogram of undiscovered recoverable oil and gas resources in major countries (regions) worldwide

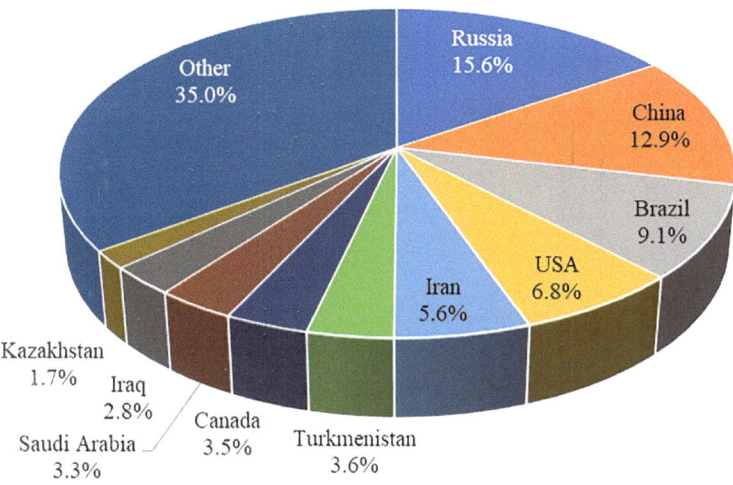

Fig. 3.16   Pie chart of undiscovered recoverable oil and gas resources in major countries (regions) worldwide

3.1.3.2   Distribution of Undiscovered Recoverable Resources in Major Basins

The undiscovered recoverable oil and gas resources worldwide are mainly distributed in 91 basins, including the Arabian Basin, West Siberian Basin, Zagros Basin, Santos Basin, Amu Darya Basin, Campos Basin, Gulf of Mexico Deepwater Basin, East Siberian Basin, and East Barents Sea Basin. The resource potentials of the Arabian Basin, West Siberian Basin, and Zagros Basin rank among the top three, and the undiscovered recoverable oil and gas resources in these three basins account for 24% of the total undiscovered recoverable resources worldwide (Figs. 3.17 and 3.18).

The undiscovered oil and gas resources in the Arabian Basin are 415.4×10^8 tons of oil equivalent, of which oil accounts for 49.3%, condensate accounts for 7.2%, and gas accounts for 43.5%. The undiscovered oil and gas resources in the West Siberian Basin are 261.1×10^8 tons of oil equivalent, of which oil accounts for 35.3%, condensate accounts for 7.2%, and gas accounts for 57.5%. The undiscovered oil and gas resources in the Zagros Basin are 236.7×10^8 tons of oil equivalent, of which oil accounts for 35.1%, condensate accounts for 5.9%, and gas accounts for 59.0%.

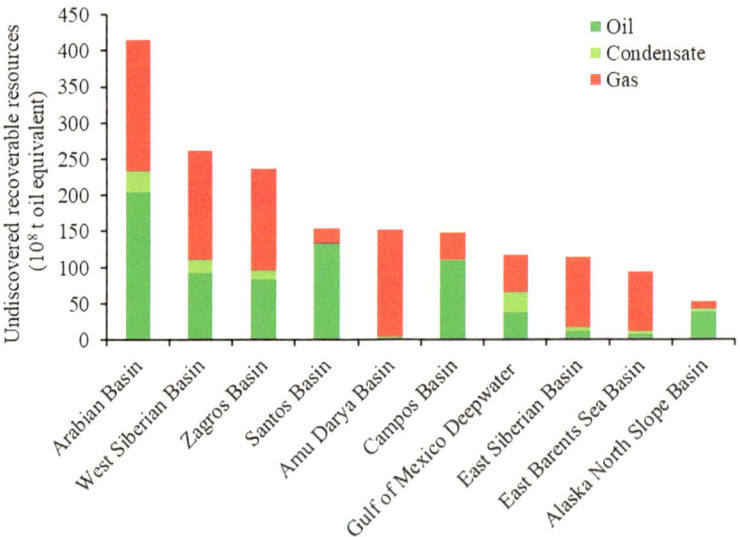

Fig. 3.17 Histogram of undiscovered recoverable oil and gas resources in major basins worldwide

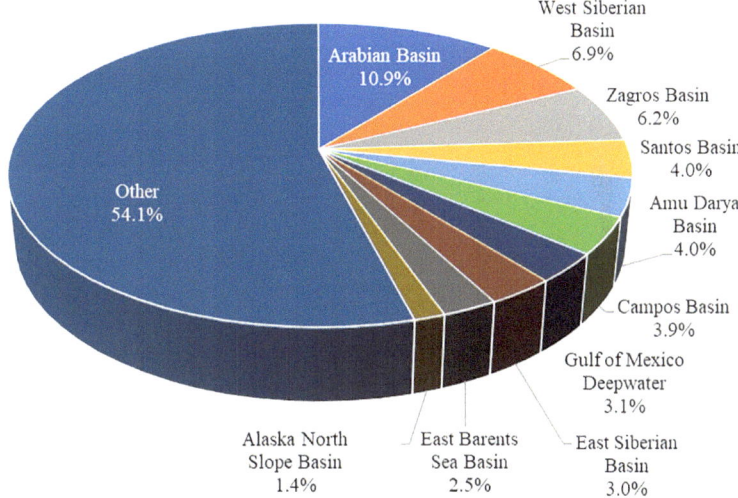

Fig. 3.18 Pie chart of undiscovered recoverable oil and gas resources in major basins worldwide

3.1.3.3 Distribution of Undiscovered Recoverable Resources in Onshore and Offshore Areas

The undiscovered recoverable oil and gas resources in onshore and offshore areas are 2268.0×10^8 and 1531.2×10^8 tons of oil equivalent, respectively, accounting for 59.7% and 40.3% of the total undiscovered recoverable resources worldwide (Fig. 3.19). Onshore conventional oil and gas resources still have huge exploration potential, and offshore conventional oil and gas resources are more important area of reserve growth in the future.

3.1.3.4 Distribution of Undiscovered Recoverable Resources in Different Types of Basins

The undiscovered recoverable oil and gas resources worldwide are mainly distributed in three types of basins, namely, passive margin basins, foreland basins, and rift basins. The undiscovered recoverable oil and gas resources distributed in these three types of basins account for 42.0%, 31.6% and 15.6% of the total undiscovered recoverable resources worldwide, respectively. The proportions of undiscovered recoverable resources in cratonic, back-arc and fore-arc basins are relatively small, and the undiscovered recoverable oil and gas resources in these three types of basins account for only 10.8% of the total undiscovered recoverable resources worldwide (Fig. 3.20).

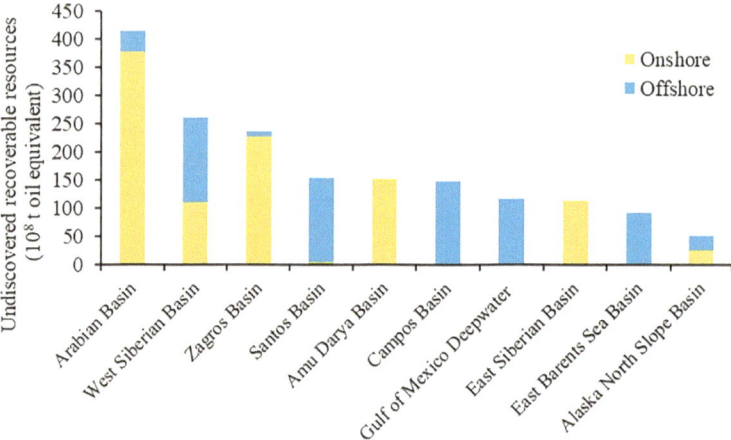

Fig. 3.19 Histogram of undiscovered recoverable oil and gas resources onshore and offshore in major basins worldwide

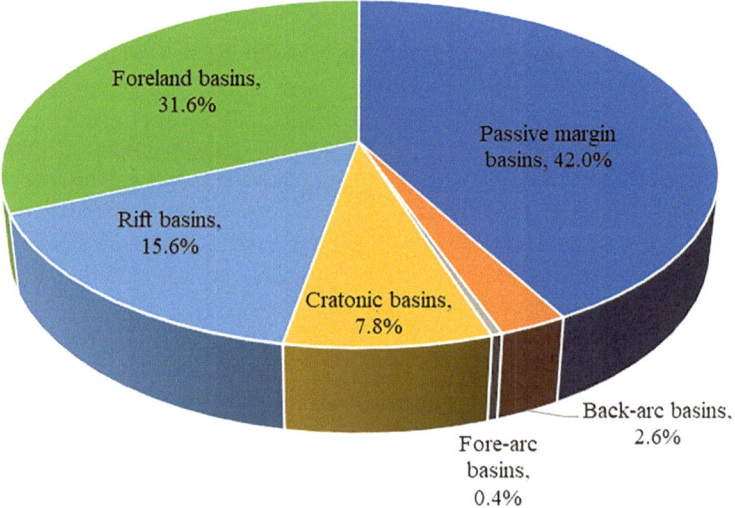

Fig. 3.20 Pie chart of the distribution of undiscovered recoverable oil and gas resources in different types of basins

The undiscovered recoverable oil and gas resources in passive margin basins are mainly distributed in the Santos Basin, Campos Basin, Gulf of Mexico Deepwater Basin, East Barents Sea Basin, and Somali Deepwater Basin. The undiscovered recoverable oil and gas resources foreland basins are mainly distributed in Arabian Basin, Zagros Basin, Eastern Venezuela Basin, South Caspian Basin, and Timan-Pechora Basin. The undiscovered recoverable oil and gas resources in rift basins are mainly distributed in West Siberian Basin, Amu Darya Basin, East African Rift System, Sirte Basin, and North Sea Basin.

3.1.3.5   Distribution of Undiscovered Recoverable Resources in Various Formations

The undiscovered recoverable oil and gas resources worldwide are distributed in the Cretaceous, Jurassic, Permian, Paleogene, and Neogene formations (these formations are arranged in descending order depending on the volumes of oil and gas resources contained therein). The undiscovered recoverable oil and gas resources distributed in these formations account for 40.3%, 16.0%, 11.2%, 11.0% and 10.1% of the world's total undiscovered recoverable resources, respectively. The undiscovered recoverable oil and gas resources in other formations have relatively low resource potential, accounting for 11.4% of the world's total undiscovered recoverable resources (Fig. 3.21).

The undiscovered recoverable oil and gas resources in the Cretaceous formations are mainly distributed in the West Siberian Basin, Santos Basin, Campos Basin, Arabian Basin, and Zagros Basin. The undiscovered recoverable oil and gas resources in the Jurassic formations are mainly distributed in the Amu Darya Basin, East Barents Sea Basin, Arabian Basin, West Siberian Basin, and Vøring Basin. The undiscovered recoverable oil and gas resources in the Permian formations are mainly distributed in Arabian Basin, Zagros Basin, Timan-Pechora Basin, North Sea Basin, and Lena-Vilyuy Basin. The undiscovered recoverable oil and gas resources in the Precambrian formations are mainly distributed in the Precambrian reservoirs in the East Siberian Basin, Sichuan Basin, and some other basins.

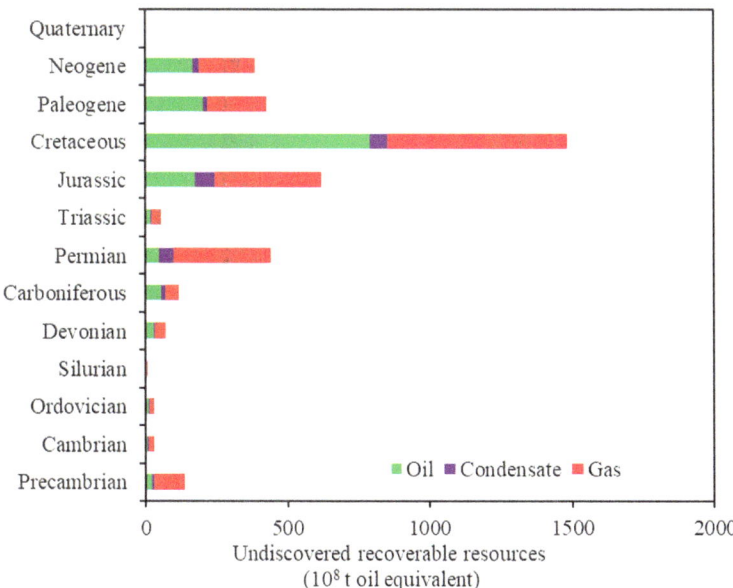

Fig. 3.21 Distribution of undiscovered recoverable oil and gas resources in geological age worldwide

3.2 Unconventional Oil and Gas Resources

The total technically recoverable unconventional resources is 7249.9×10^8 tons of oil equivalent, of which the technically recoverable unconventional oil resources are 4244.9×10^8 tons, accounting for 58.6% of the total unconventional oil and gas resources, and the technically recoverable unconventional gas resources are 351.6×10^{12} m^3, accounting for 41.4% of the total unconventional oil and gas resources.

Among unconventional oil resources worldwide, the amounts of recoverable oil shale resources are the largest, reaching 1540.8×10^8 tons, accounting for 36.3%; the recoverable heavy oil resources are 1274.8×10^8 tons, accounting for 30.3%; the recoverable shale oil resources are 790.4×10^8 tons, accounting for 18.6%; and the recoverable oil sands are 638.9×10^8 tons, accounting for 15.1% (Figs. 3.22 and 3.23).

Among unconventional gas resources, the volume of shale gas is the largest. The volume of technically recoverable shale gas is 235.9×10^{12} m^3, accounting for 67.1% of the world's total recoverable gas resources; the volume of technically recoverable CBM are 53×10^{12} m^3, accounting for 15.1%; and the volume of technically recoverable tight gas are 62.7×10^{12} m^3, accounting for 17.8% (Figs. 3.24 and 3.25).

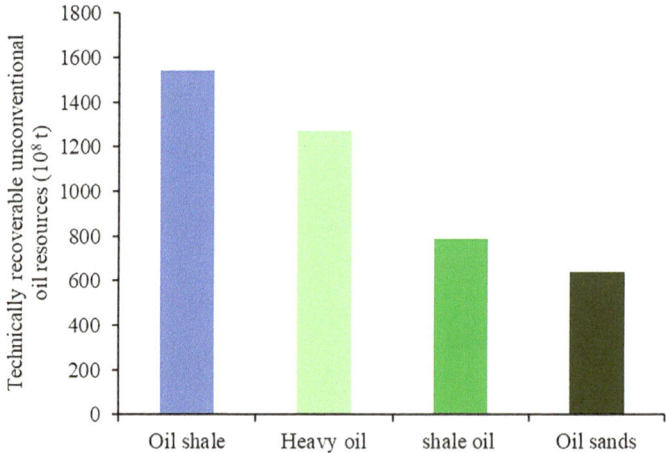

Fig. 3.22 Histogram of technically recoverable unconventional oil resources worldwide

Fig. 3.23 Pie chart of technically recoverable unconventional oil resources worldwide

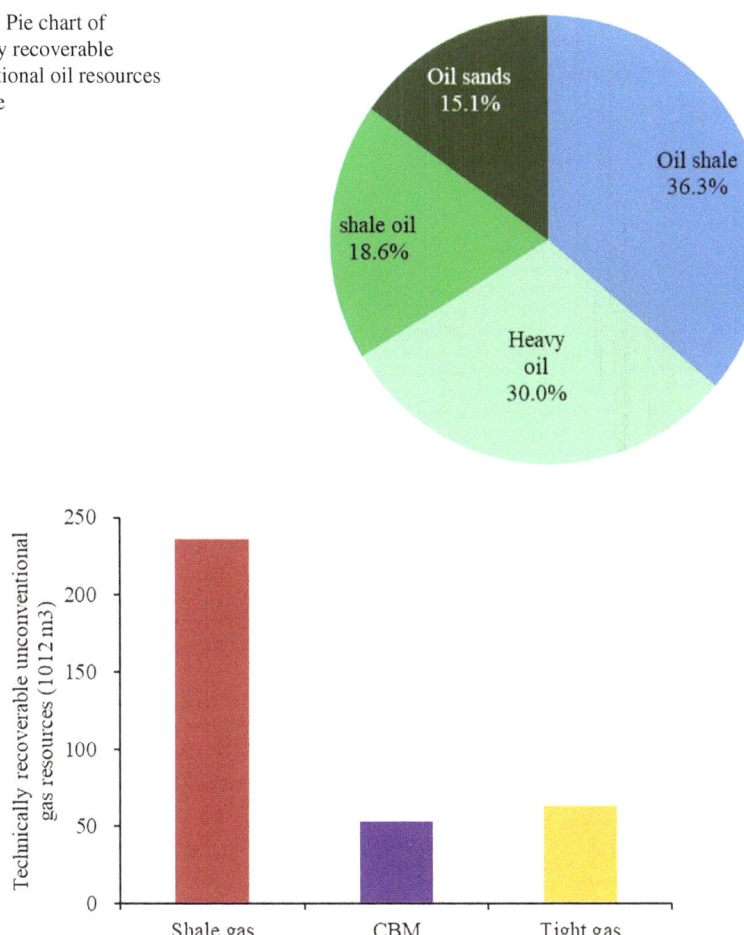

Fig. 3.24 Histogram of technically recoverable unconventional gas resources worldwide

Fig. 3.25 Pie chart of
technically recoverable
unconventional gas resources
worldwide

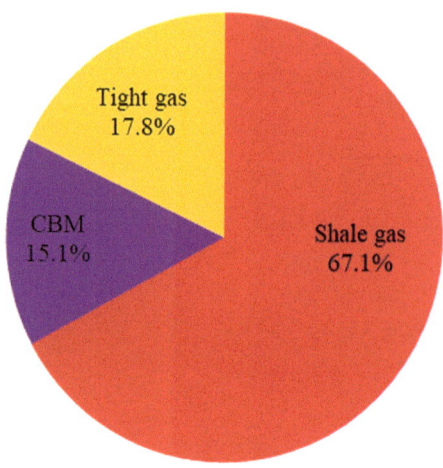

3.2.1   Distribution of Recoverable Unconventional Oil and Gas Resources in Various Regions

3.2.1.1   Distribution of Recoverable Unconventional Oil Resources in Various Regions

69.7% of the world's recoverable unconventional oil resources are concentrated in North America, Russia, and Central and South America. The technically recoverable unconventional oil resources in North America are 1586.2×10^8 tons, accounting for 37.4% of the world's total recoverable unconventional oil resources and mainly including oil shale, oil sands, and heavy oil; the technically recoverable unconventional oil resources in Russia are 683.1×10^8 tons, mainly including oil shale, shale oil, and oil sands; and the technically recoverable unconventional oil resources in Central and South America are 660.8×10^8 tons, mainly including heavy oil and oil shale (Figs. 3.26 and 3.27).

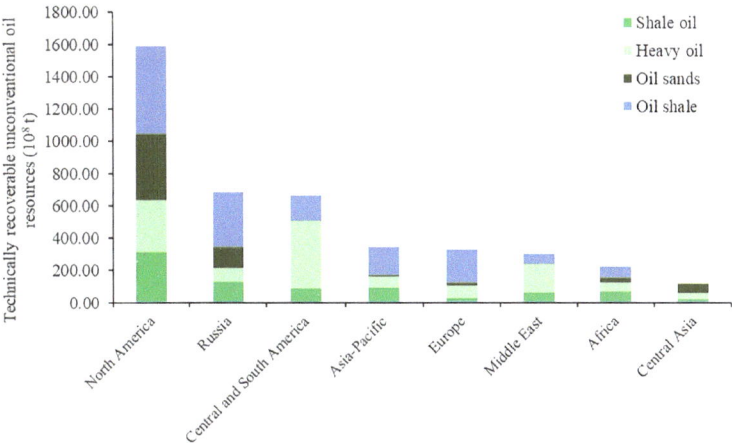

Fig. 3.26 Histogram of the distribution of technically recoverable unconventional oil resources in various regions worldwide

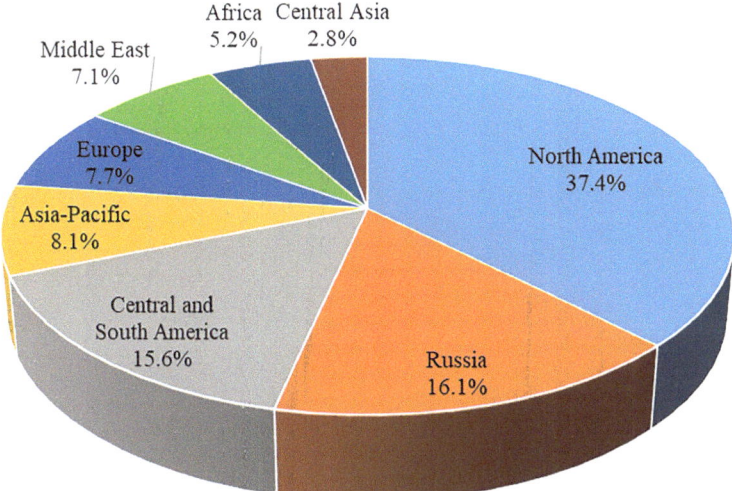

Fig. 3.27 Pie chart of the distribution of technically recoverable unconventional oil resources in various regions worldwide

3.2.1.2 Distribution of Recoverable Unconventional Gas Resources in Various Regions

60.4% of the world's recoverable unconventional gas resources are concentrated in North America, Asia–Pacific, Central and South America. The recoverable unconventional gas resources in North America are 99.9×10^{12} m^3, accounting for 28.4% of the world's total recoverable unconventional gas resources and mainly including shale gas and CBM; the recoverable unconventional gas resources in Asia–Pacific are 61.9×10^{12} m^3, mainly including shale gas; and the recoverable unconventional gas resources in Central and South America are 50.7×10^{12} m^3, mainly including shale gas and tight gas (Figs. 3.28 and 3.29).

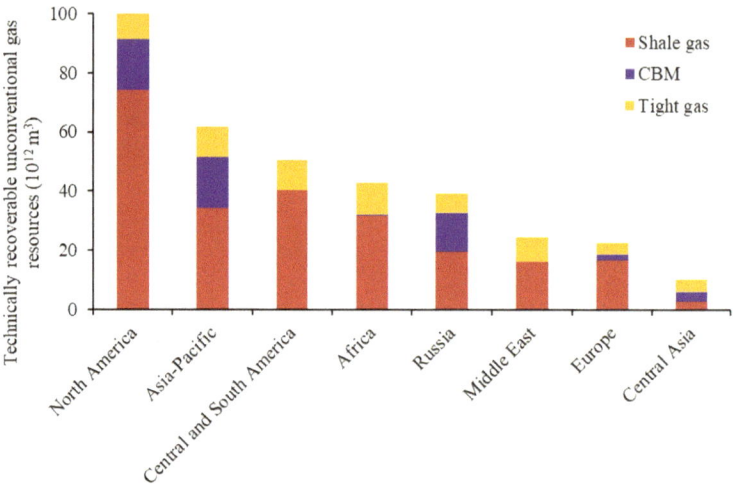

Fig. 3.28 Histogram of the distribution of technically recoverable unconventional gas resources in various regions worldwide

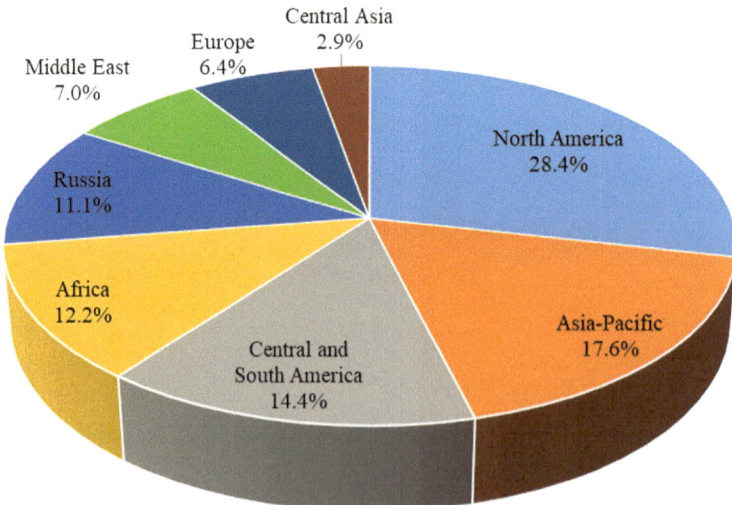

Fig. 3.29 Pie chart of the distribution of technically recoverable unconventional gas resources in various regions worldwide

3.2.2 Distribution of Recoverable Unconventional Oil and Gas Resources in Major Countries (Regions)

3.2.2.1 Distribution of Recoverable Unconventional Oil Resources in Major Countries (Regions)

Unconventional oil resources worldwide are distributed in 52 countries (regions), and more than 63.8% of the world's recoverable unconventional oil resources are concentrated in the USA, Russia, Canada, Venezuela, Saudi Arabia et al. The recoverable unconventional oil resources in the top three countries in terms of the volumes of such resources, such as the USA, account for 52.4% of the world's total recoverable unconventional oil resources. The recoverable unconventional oil resources in the USA are 1024.3×10^8 tons, accounting for 24.1% of the world's total recoverable unconventional oil resources and mainly including oil shale, heavy oil, and shale oil; the recoverable unconventional oil resources in Russia are 683.1×10^8 tons, accounting for 16.1% and mainly including oil shale, oil sands, and shale oil; and the recoverable unconventional oil resources in Canada are 413.9×10^8 tons, accounting for 9.8% and mainly including oil sands (Figs. 3.30 and 3.31).

3.2.2.2 Distribution of Recoverable Unconventional Gas Resources in Major Countries (Regions)

Unconventional gas resources worldwide are distributed in 35 countries, and about 74.4% of the world's recoverable unconventional gas resources are concentrated in 10

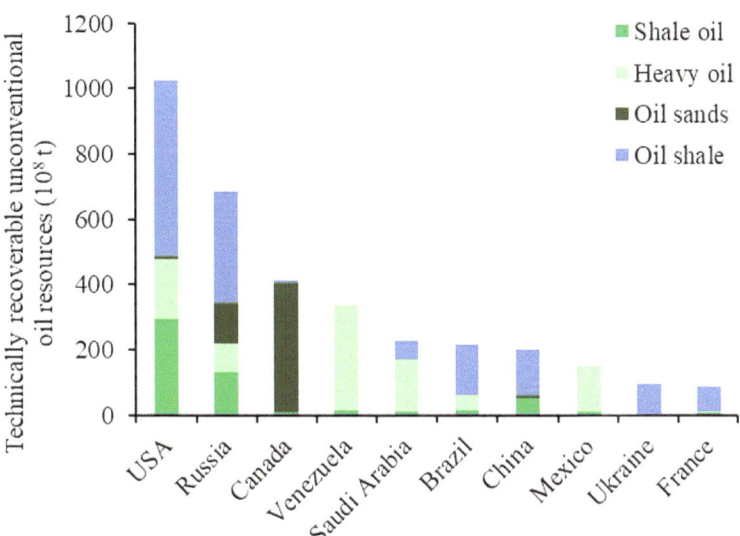

Fig. 3.30 Histogram of technically recoverable unconventional oil resources distributed in major countries (regions) worldwide

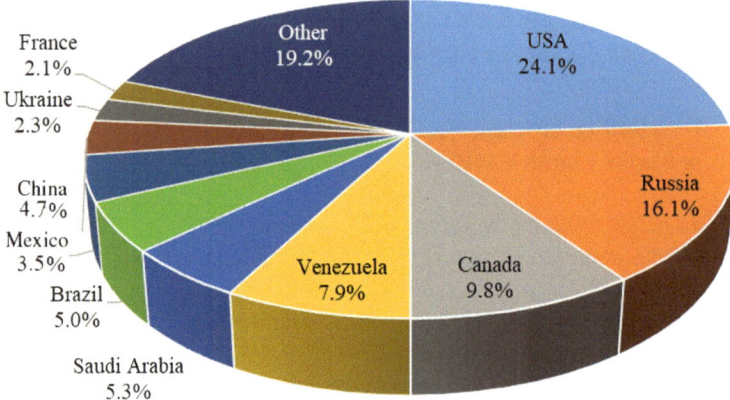

Fig. 3.31 Pie chart of technically recoverable unconventional oil resources distributed in major countries (regions) worldwide

countries, including the USA, Russia, China, Canada, Argentina, Algeria, Australia, Saudi Arabia, Brazil, and Indonesia. The recoverable unconventional gas resources in the top three countries in terms of the volumes of such resources, such as USA, Russia, and China, account for 41.0% of the world's total recoverable unconventional gas resources. The recoverable unconventional gas resources in USA are 71.5 $\times$ 10^{12} m^3, accounting for 20.3% of the world's total recoverable unconventional gas resources, and the volume of recoverable shale gas accounts for 81.6% of the total recoverable unconventional gas resources in the USA. The recoverable unconventional gas resources in Russia are 39.2 $\times$ 10^{12} m^3, accounting for 11.1% of the world's total recoverable unconventional gas resources and mainly including shale gas and CBM. The recoverable unconventional gas resources in China are 33.5 $\times$ 10^{12} m^3, accounting for 9.5% of the world's total recoverable unconventional gas resources and mainly including shale gas, CBM and tight gas (Figs. 3.32 and 3.33).

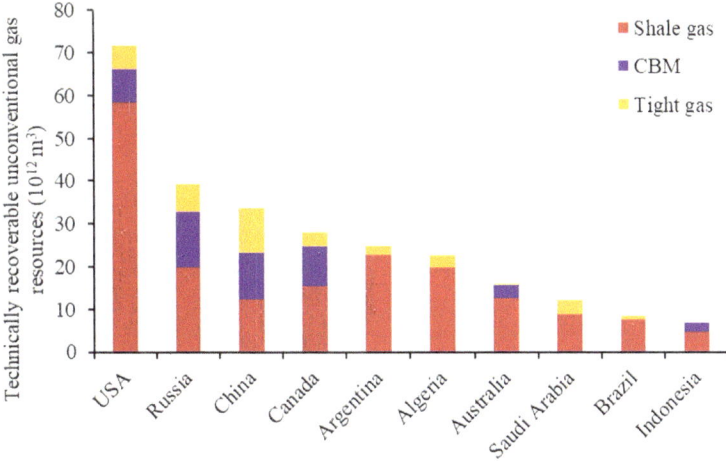

Fig. 3.32 Histogram of technically recoverable unconventional gas resources distributed in major countries (regions) worldwide

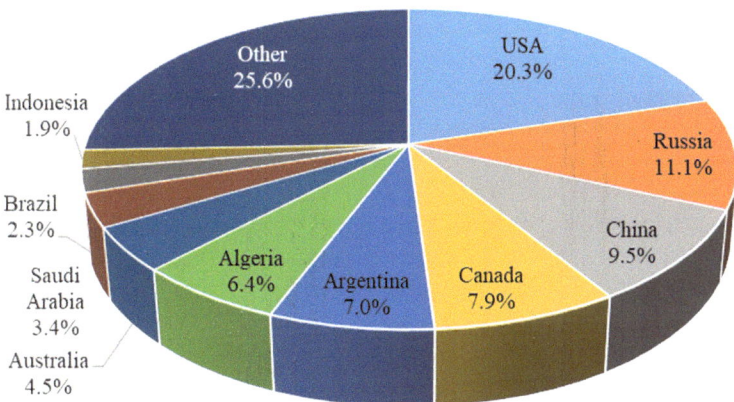

Fig. 3.33 Pie chart of technically recoverable unconventional gas resources distributed in major countries (regions) worldwide

3.2.3 Distribution of Recoverable Unconventional Oil and Gas Resources in Major Basins

3.2.3.1 Distribution of Recoverable Unconventional Oil Resources in Major Basins

Unconventional oil resources worldwide are mainly distributed in 136 basins, and 70% of these resources are concentrated in 17 basins, including the Alberta Basin, the Eastern Venezuela Basin, the Arabian Basin, the Uinta Basin in the USA, and

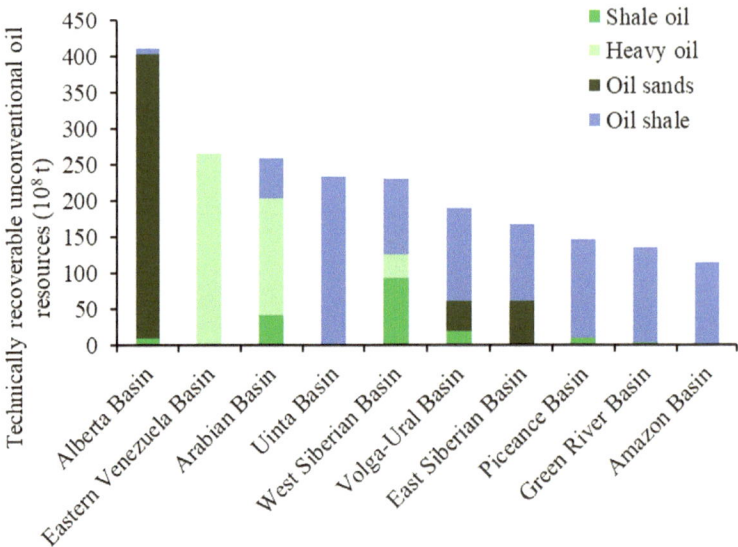

Fig. 3.34 Histogram of technically recoverable unconventional oil resources distributed in major basins worldwide

the West Siberian Basin. The recoverable unconventional oil resources in the top three basins in terms of the volumes of such resources, such as the Alberta Basin, account for 22.1% of the world's total recoverable unconventional oil resources. The recoverable unconventional oil resources in the Alberta Basin are 411.3×10^8 tons, accounting for 9.7% of the world's total recoverable unconventional oil resources and mainly including oil sands and oil shale; the recoverable unconventional oil resources in the Eastern Venezuela Basin are 266.5×10^8 tons, accounting for 6.3%; and the recoverable unconventional oil resources in the Arabian Basin are 260.0×10^8 tons, accounting for 6.1% and mainly including heavy oil, oil shale, and shale oil (Figs. 3.34 and 3.35).

3.2.3.2 Distribution of Recoverable Unconventional Gas Resources in Major Basins

Unconventional gas resources worldwide are mainly distributed in 86 basins, and 80% of these resources are concentrated in 26 basins, including the Alberta Basin, Gulf Coast Basin, Zagros Basin, Appalachian Basin, and Neuquén Basin. The recoverable unconventional gas resources in the top three basins in terms of the volumes of such resources, such as the Alberta Basin, account for 24.5% of the world's total recoverable unconventional gas resources. The recoverable unconventional gas resources in the Alberta Basin are 27.6×10^{12} m^3, accounting for 7.9% of the world's total recoverable unconventional gas resources and mainly including shale gas and CBM; the recoverable unconventional gas resources in the Gulf Coast Basin are 26.8

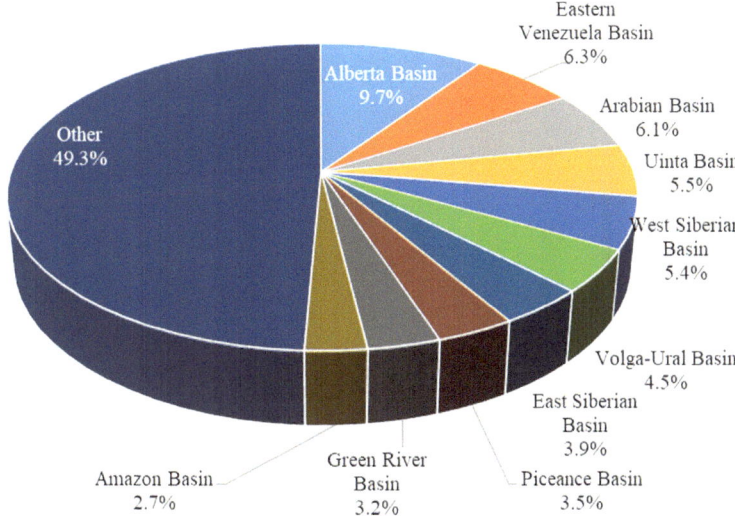

Fig. 3.35 Pie chart of technically recoverable unconventional oil resources distributed in major basins worldwide

$\times 10^{12}$ m^3, accounting for 7.6% and mainly including shale gas; and the recoverable unconventional gas resources in the Appalachian Basin are 20.7×10^{12} m^3, accounting for 5.9% and mainly including shale gas and tight gas (Figs. 3.36 and 3.37).

3.2.4 Distribution of Recoverable Unconventional Oil and Gas Resources in Different Types of Basins

3.2.4.1 Distribution of Recoverable Unconventional Oil Resources in Different Types of Basins

The technically recoverable unconventional oil resources worldwide are mainly distributed in three types of basins, namely, foreland basins, cratonic basins, and rift basins. The technically recoverable unconventional oil resources in these three types of basins account for 64.4%, 16.2% and 10.2% of the world's total recoverable unconventional oil resources, respectively (Fig. 3.38). The recoverable unconventional oil resources concentrated in foreland basins mainly include heavy oil, oil shale and shale oil, which account for 33%, 30.7% and 17.4%, respectively. The recoverable unconventional oil resources concentrated in cratonic basins mainly include oil shale, oil sands and shale oil, which account for 71.3%, 14.6% and 14.1%, respectively. The recoverable unconventional oil resources concentrated in rift basins

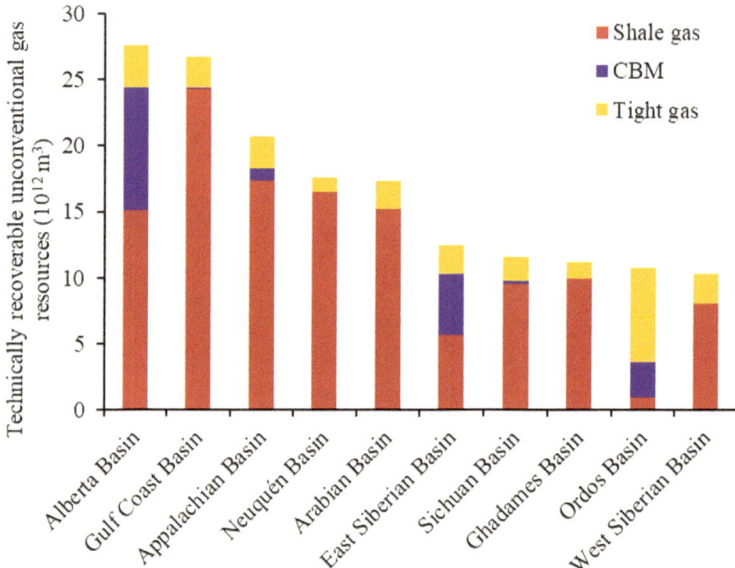

Fig. 3.36 Histogram of technically recoverable unconventional gas resources distributed in major basins worldwide

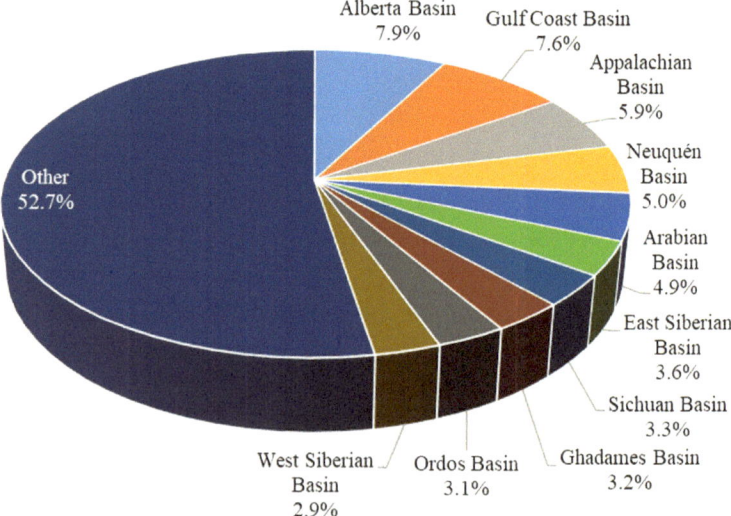

Fig. 3.37 Pie chart of technically recoverable unconventional gas resources distributed in major basins worldwide

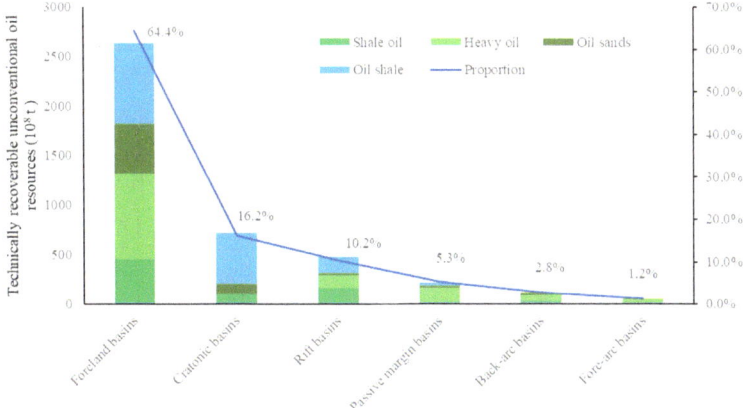

Fig. 3.38 Histogram of technically recoverable unconventional oil resources distributed in different types of basins worldwide

mainly include shale oil, heavy oil and oil shale, which account for 38.2%, 32.6% and 26.3%, respectively.

3.2.4.2 Distribution of Recoverable Unconventional Gas Resources in Different Types of Basins

The technically recoverable unconventional gas resources worldwide are mainly distributed in three types of basins, namely, foreland basins, cratonic basins, and rift basins. The technically recoverable unconventional gas resources in these three types of basins account for 56.9%, 19.6% and 8.5% of the world's total recoverable unconventional gas resources, respectively (Fig. 3.39). Shale gas and CBM, especially shale gas resources, are mainly concentrated in foreland basins, accounting for 85.2% and 10.9%, respectively. Shale gas and CBM concentrated in cratonic basins account for 85.9% and 13.9%, respectively. Shale gas and CBM concentrated in rift basins account for 67.3 and 30.6%, respectively. Tight gas is mainly concentrated in foreland basins.

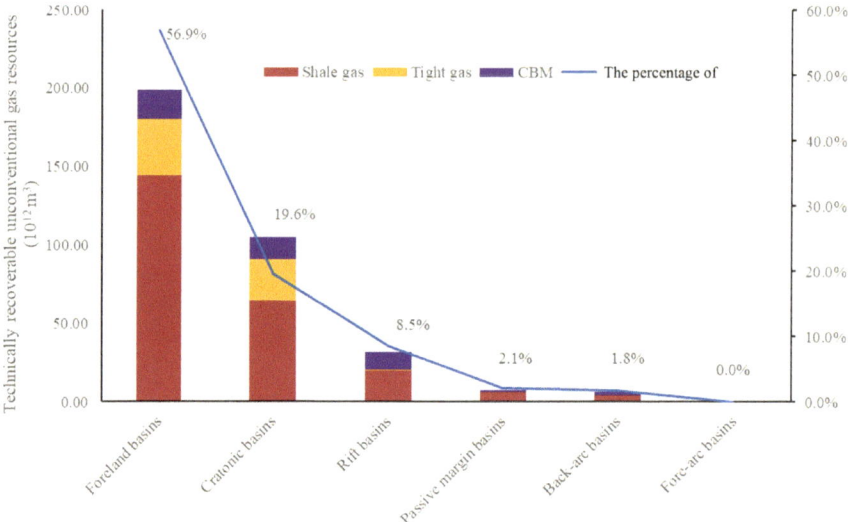

Fig. 3.39 Histogram of technically recoverable unconventional gas resources distributed in different types of basins worldwide

3.2.5 Distribution of Recoverable Unconventional Oil and Gas Resources in Geological Age

3.2.5.1 Distribution of Recoverable Unconventional Oil Resources in Geological Age

The technically recoverable unconventional oil resources are mainly distributed in the Cretaceous, Jurassic, Paleogene, Neogene, and Carboniferous formations. The recoverable unconventional oil resources in these formations account for 32%, 15.8%, 15.7%, 12.5% and 7% of the world's total recoverable unconventional oil resources, respectively. Other formations have relatively small resource potential, and the recoverable unconventional oil resources in other formations account for only 17% of the world's total recoverable unconventional oil resources (Fig. 3.40).

The technically recoverable shale oil resources are mainly distributed in the Cretaceous, Jurassic, Carboniferous, Devonian, and Silurian formations (in descending order depending on the volume of shale oil contained therein), and the recoverable shale oil resources in these formations account for 29.8%, 22.9%, 15.4%, 14.3% and 5.1%, respectively. The technically recoverable heavy oil resources are mainly concentrated in the Neogene, Cretaceous, Paleogene, Jurassic, and Triassic formations, and the recoverable heavy oil resources in these formations account for 33.7%, 24.9%, 23.2%, 10.4% and 5.5%, respectively. The technically recoverable oil sands are mainly concentrated in the Cretaceous, Neogene, Cambrian, and Carboniferous formations, and the recoverable oil sands in these formations account for 69%, 10.3%,

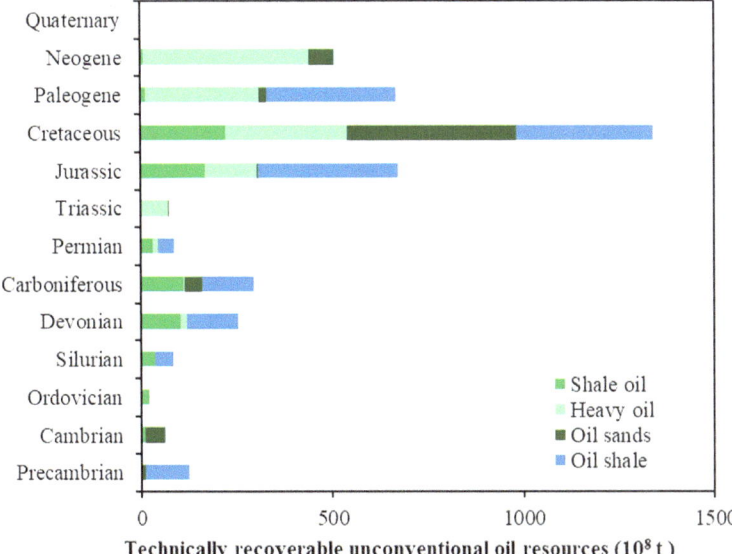

Fig. 3.40 Histogram of technically recoverable unconventional oil resources in geological age worldwide

8% and 7.2%, respectively. The technically recoverable oil shale are mainly concentrated in the Jurassic, Cretaceous, Paleogene, Carboniferous, and Devonian formations, and the recoverable oil shale in these formations account for 23.7%, 23.3%, 21.8%, 8.8% and 8.7%, respectively.

3.2.5.2 Distribution of Recoverable Unconventional Gas Resources in geological age

The technically recoverable unconventional gas resources worldwide are mainly distributed in the Cretaceous, Jurassic, Devonian, Silurian, and Carboniferous formations (in descending order depending on the volumes of such resources contained therein), and the recoverable unconventional gas resources in these formations account for 29.9%, 19.8%, 17.3%, 9.8% and 9.4% of the world's total recoverable unconventional gas resources, respectively. Other formations have relatively small resource potential, and the recoverable unconventional gas resources in other formations account for only 13.8% of the world's total recoverable unconventional gas resources (Fig. 3.41).

The technically recoverable shale gas resources are mainly distributed are mainly enriched in the Cretaceous, Devonian, Jurassic, Silurian, and Carboniferous formations (in descending order depending on the volumes of such resources contained therein), and the recoverable shale gas resources in these formations account for 30.9%, 20.6%, 18.9%, 10.6% and 6%, respectively. The technically recoverable tight

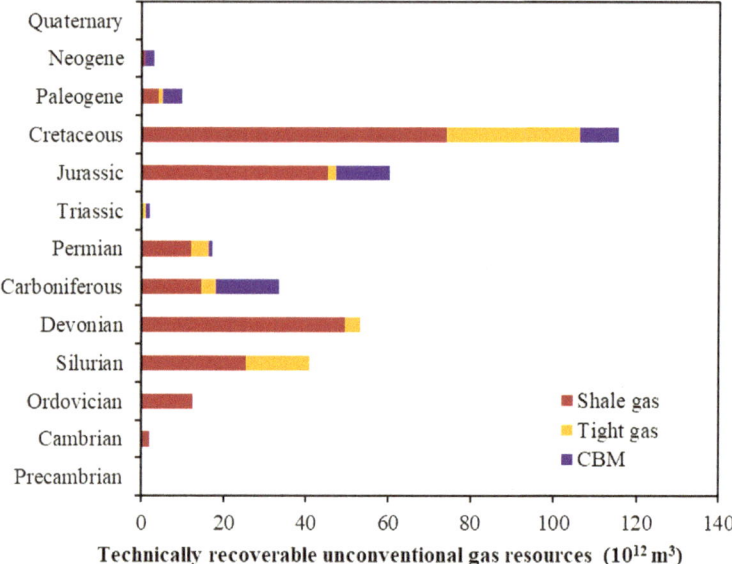

Fig. 3.41 Histogram of technically recoverable unconventional gas resources in the percentage of worldwide

gas resources are mainly concentrated in the Cretaceous, Silurian, Permian, Devonian, and Carboniferous formations, and the recoverable tight gas resources in these formations account for 50.9%, 24.6%, 6.5%, 6.0% and 5.7%, respectively. The technically recoverable CBM resources are mainly concentrated in the Carboniferous, Jurassic, Cretaceous, Paleogene, and Neogene formations, and the recoverable CBM resources in these formations account for 32.4%, 27.8%, 20.1%, 9.9% and 5.2%, respectively.

Chapter 4
Distribution of Oil and Gas Resources in North America

North America includes the USA, Canada, Greenland, Mexico, and other countries and regions, covering an area of 2177.9×10^4 km^2, where 82 sedimentary basins have developed. The onshore sedimentary zone covers an area of 1986.6×10^4 km^2 and is dominated by foreland and cratonic basins. The offshore sedimentary zone covers an area of 191.3×10^4 km^2 and is dominated by passive margin basins. Twenty percent of the world's total oil and gas resources are concentrated in North America. The total recoverable oil and gas resources in North America are 3562.3×10^8 t of oil equivalent.

4.1 Conventional Oil and Gas Resources

The conventional oil and gas resources in North America are 1122.2×10^8 t of oil equivalent, accounting for 9.8% of the world's total conventional resources. The recoverable oil and gas reserves in North America are 586.0×10^8 t of oil equivalent, accounting for 9.0% of the global recoverable reserves; the remaining recoverable oil and gas reserves are 156.1×10^8 t of oil equivalent, accounting for 4.0% of the world's total remaining recoverable reserves; the estimated reserve growth from discovered oil and gas fields in the future are 110.0×10^8 t of oil equivalent, accounting for 9.8% of the world's total reserve growth; and the undiscovered recoverable oil and gas resources are 426.2×10^8 t of oil equivalent, accounting for 11.2% of the world's total undiscovered recoverable resources.

© The Author(s) 2024
L. Dou et al., *Global Oil and Gas Resources: Potential and Distribution*,
https://doi.org/10.1007/978-981-97-4756-6_4

4.1.1   Distribution of Remaining Recoverable Reserves

The remaining recoverable oil and gas reserves in North America are 156.1×10^8 tons of oil equivalent, including 89.5×10^8 t of oil, 5.3×10^8 tons of condensate, and 7.2×10^{12} m^3 of gas.

4.1.1.1   Distribution of Remaining Recoverable Reserves in Major Countries

The remaining recoverable oil and gas reserves in the USA are 111.8×10^8 t of oil equivalent, of which the remaining recoverable oil accounts for 59.6%, the remaining recoverable condensate accounts for 2.5%, and the remaining recoverable gas accounts for 37.9%. The remaining recoverable oil and gas reserves in Mexico rank second, reaching 24.8×10^8 t of oil equivalent, of which the remaining recoverable accounts for 63.6%, the remaining recoverable condensate accounts for 6.9%, and the remaining recoverable gas accounts for 29.5%. The remaining recoverable oil and gas reserves in Canada are 19.5×10^8 t of oil equivalent, of which the remaining recoverable oil accounts for 36.0%, the remaining recoverable condensate accounts for 4.1%, and the remaining recoverable gas accounts for 59.9% (Figs. 4.1 and 4.2).

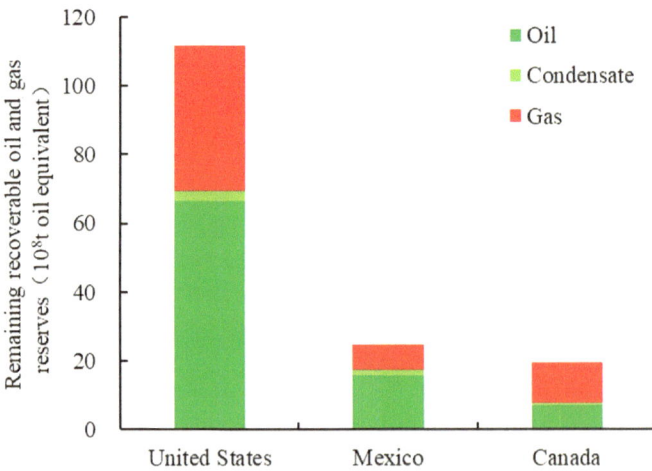

Fig. 4.1 Histogram of remaining recoverable oil and gas reserves in major North American countries

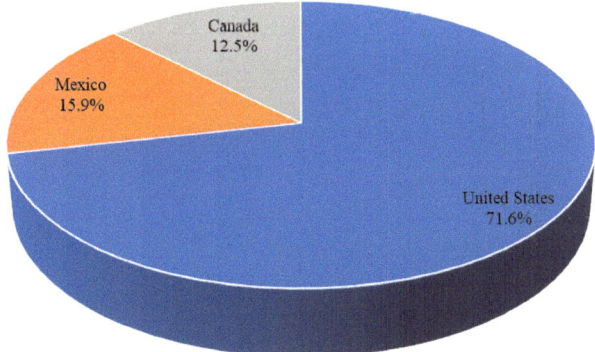

Fig. 4.2 Pie chart of remaining recoverable oil and gas reserves in major North American countries

4.1.1.2 Distribution of Remaining Recoverable Reserves in Major Basins

The remaining recoverable oil and gas reserves in North America are mainly concentrated in the Alaska North Slope Basin, Gulf of Mexico Deepwater Basin, Gulf Coast Basin, Sureste Basin, Permain Basin, Saint Joaquin Basin, North Atlantic Basin, etc. The remaining recoverable oil and gas reserves in the top three basins account for 51.3% of the total remaining recoverable reserves in North America (Figs. 4.3 and 4.4).

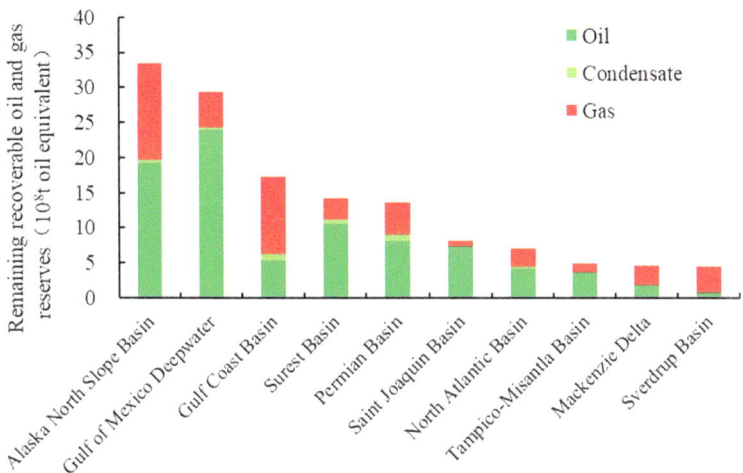

Fig. 4.3 Histogram of the remaining recoverable oil and gas reserves in major basins in North America

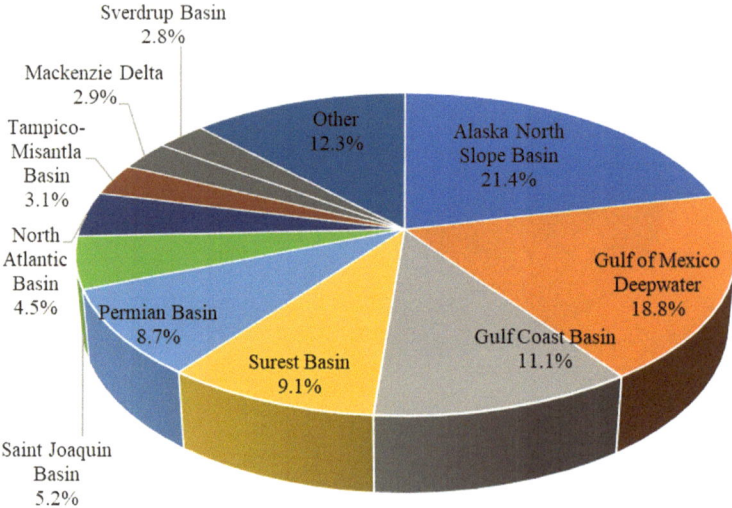

Fig. 4.4 Pie chart of the remaining recoverable oil and gas reserves in major basins in North America

The remaining recoverable oil and gas reserves in the Alaska North Slope Basin are 33.5×10^8 t of oil equivalent, of which oil accounts for 57.5%, condensate accounts for 1.5%, and gas accounts for 41.0%. The remaining recoverable oil and gas reserves in the Gulf of Mexico Deepwater Basin rank second and are 29.3×10^8 t of oil equivalent, of which oil accounts for 81.6%, condensate accounts for 1.4%, and gas accounts for 17%. The remaining recoverable oil and gas reserves in the Gulf Coast Basin are 17.3×10^8 t of oil equivalent, of which oil accounts for 31.1%, condensate accounts for 5.3%, and gas accounts for 63.6%.

The remaining recoverable oil and gas reserves in onshore areas are 86.7×10^8 t of oil equivalent, of which oil accounts for 42.2%, condensate accounts for 2.1%, and gas accounts for 55.7%. The remaining recoverable oil and gas reserves in offshore areas are 69.4×10^8 t of oil equivalent, of which oil accounts for 64.9%, condensate accounts for 2.9%, and gas accounts for 32.2%.

4.1.1.3 Distribution of Remaining Recoverable Reserves in Onshore and Offshore Areas

The remaining recoverable reserves of onshore and offshore oil and gas are similar, accounting for 55.5% and 44.5%, respectively. The volumes of remaining recoverable oil reserves in offshore areas are larger than those of remaining recoverable gas reserves in such areas. The volumes of remaining recoverable gas reserves in onshore areas are larger than those of remaining recoverable oil reserves in such areas (Fig. 4.5).

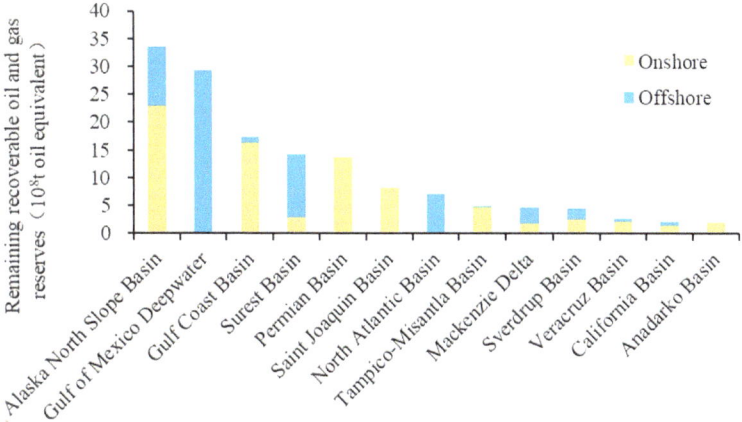

Fig. 4.5 Histogram of the remaining recoverable reserves onshore and offshore in major basins in North America

4.1.2   Trend of Reserve Growth from Discovered Oil and Gas Fields

The total reserve growth from discovered oil and gas fields in North America is 110.0 $\times 10^8$ t of oil equivalent, including 69.8 $\times 10^8$ t of oil, which accounts for 63.4%, 4.1 $\times 10^8$ t of oil, which accounts for 3.7%, and 4.2 $\times 10^{12}$ m^3 of gas, which accounts for 32.9%.

4.1.2.1   Distribution of Reserve Growth in Discovered Oil and Gas Fields in Major Countries

The future reserve growth in discovered oil and gas fields in the USA is the largest, accounting for 57.6% of the total reserve growth in North America, and the growth in oil reserves is greater than that in gas reserves. The reserve growth in discovered oil and gas fields in Mexico and Canada account for 29.2% and 13.2% of the total reserve growth in North America, respectively. The future growth in gas reserves in Canada is slightly greater than that in oil reserves in the same country. The future reserve growth in discovered oil and gas fields in Mexico are mainly oil reserves (Figs. 4.6 and 4.7).

The future reserve growth in discovered oil and gas fields in the USA are 63.4 $\times 10^8$ t of oil equivalent, of which oil accounts for 59.7%, condensate accounts for 5.5%, and gas accounts for 34.8%. The future reserve growth in discovered oil and gas fields in Canada are 14.5 $\times 10^8$ t of oil equivalent, of which oil accounts for 47.1%, condensate accounts for 4.2%, and gas accounts for 48.7%. The future reserve growth in discovered oil and gas fields in Mexico are 32.2 $\times 10^8$ t of oil equivalent, of which oil accounts for 78.1% and gas accounts for 21.9%.

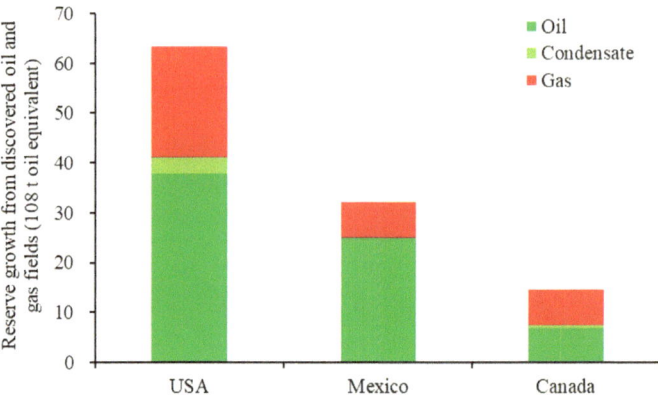

Fig. 4.6 Histogram of future reserve growth in discovered oil and gas fields in major North American countries

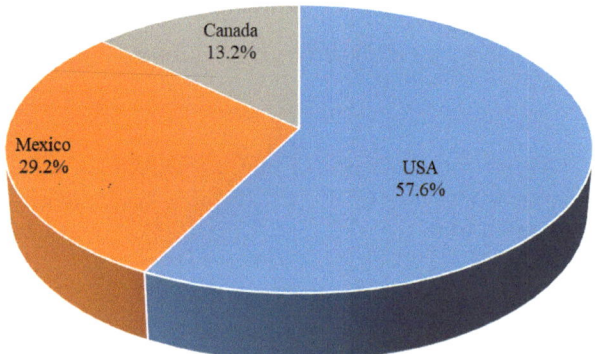

Fig. 4.7 Pie chart of future reserve growth in discovered oil and gas fields in major North American countries

4.1.2.2   Distribution of Reserve Growth in Discovered Oil and Gas Fields in Major Basins

The future reserve growth in discovered oil and gas fields in North America are mainly distributed in the Gulf of Mexico Deepwater Basin, Sureste Basin, Alaska North Slope Basin, Tampico-Misantla Basin, and Alberta Basin. The Gulf of Mexico Deepwater Basin, Sureste Basin and Alaska North Slope Basin rank among the top three basins in terms of reserve growth in the future. The future reserve growth in discovered oil and gas fields in these basins account for 55.7% of the total reserve growth in North America (Figs. 4.8 and 4.9).

The future reserve growth in discovered oil and gas fields in the Gulf of Mexico Deepwater Basin are 26.5×10^8 t of oil equivalent, of which oil accounts for 80.9%, condensate accounts for 0.9%, and gas accounts for 18.2%. The future reserve growth

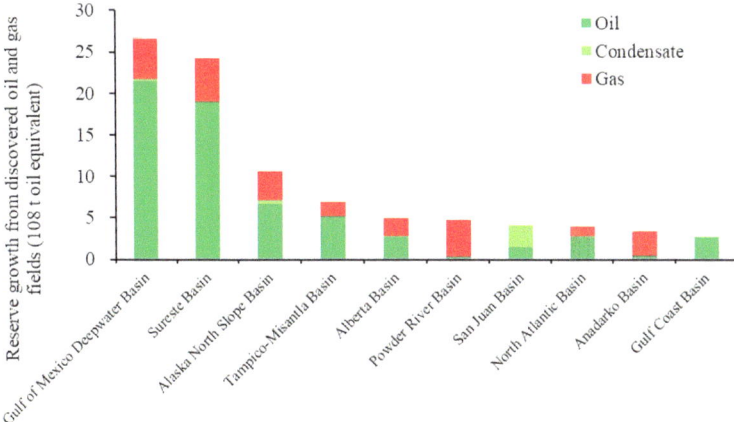

Fig. 4.8 Histogram of future reserve growth in discovered oil and gas fields in major basins in North America

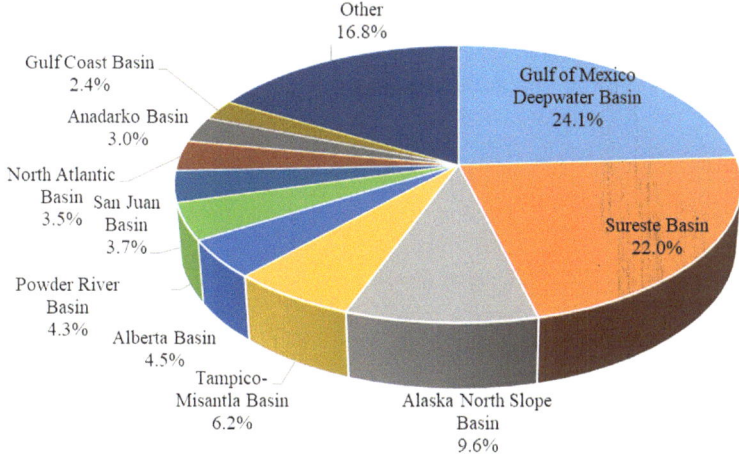

Fig. 4.9   Pie chart of future reserve growth in discovered oil and gas fields in major basins in North America

in discovered oil and gas fields in the Sureste Basin are 24.2×10^8 t of oil equivalent, of which oil accounts for 78.4% oil and gas accounts for 21.6%. The future reserve growth in discovered oil and gas fields in the Alaska North Slope Basin are 10.5×10^8 t of oil equivalent, of which oil accounts for 63.1%, condensate accounts for 3.6%, and gas accounts for 33.3%.

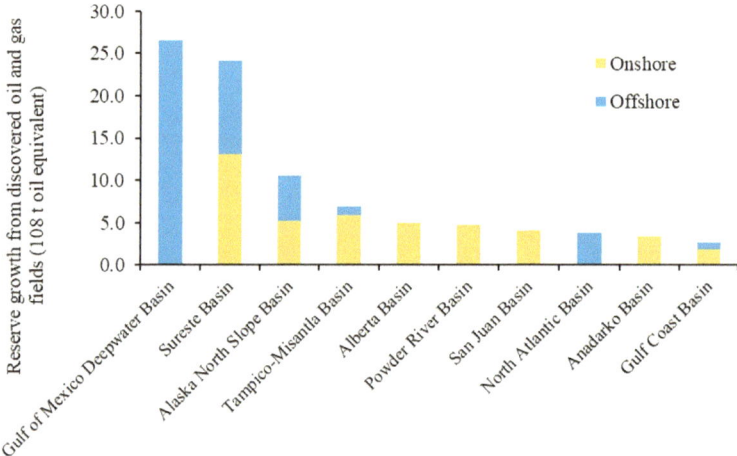

Fig. 4.10 Histogram of future reserve growth in discovered onshore and offshore oil and gas fields in major basins in North America

4.1.2.3 Distribution of Reserve Growth in Discovered Onshore and Offshore Oil and Gas Fields

The future reserve growth in discovered onshore oil and gas fields and those from discovered offshore fields account for 50.9% and 49.1%, respectively. The additional oil reserves in discovered offshore fields are three times the additional gas reserves in the same fields, while the additional oil reserves in onshore fields are close to the additional gas reserves in the same fields (Fig. 4.10). The future reserve growth in discovered onshore oil and gas fields is 56.0×10^8 t of oil equivalent, of which oil accounts for 53.5%, condensate accounts for 6.1%, and gas accounts for 40.4%. The future reserve growth in discovered onshore oil and gas fields are 54.0×10^8 t of oil equivalent, of which oil accounts for 73.6%, condensate accounts for 1.2%, and gas accounts for 25.2%.

4.1.3 Characteristics of Distributions of Undiscovered Oil and Gas Resources

The undiscovered oil and gas resources in North America are 426.2×10^8 t of oil equivalent. The amount of undiscovered oil in North America is larger than the volume of undiscovered gas. The undiscovered oil and gas resources in North America include 160.1×10^8 t of oil, accounting for 37.6%, 56.0×10^8 t of condensate, accounting for 13.1%, and 24.6×10^{12} m^3 of gas, accounting for 49.3%.

4.1.3.1 Distribution of Undiscovered Oil and Gas Resources in Major Countries

Among major North American countries, the USA has the largest volumes of undiscovered oil and gas resources, and the volume of undiscovered oil is larger than that of undiscovered gas in the USA. In Canada, the volume of undiscovered gas is two times the amount of undiscovered oil. In Mexico, the amount of undiscovered oil is three times the volume of undiscovered gas (Figs. 4.11 and 4.12).

The total undiscovered oil and gas resources in the USA is 259.0×10^8 t of oil equivalent, of which oil accounts for 37.3%, condensate accounts for 16.8%, and gas accounts for 45.9%. The total undiscovered oil and gas resources in Canada are 131.3×10^8 t of oil equivalent, of which oil accounts for 31.4%, condensate accounts for 5.5%, and gas accounts for 63.1%. The total undiscovered oil and gas

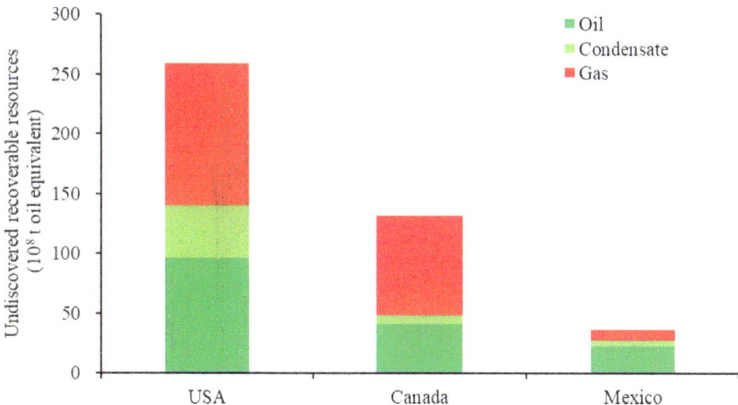

Fig. 4.11 Histogram of undiscovered recoverable oil and gas resources in major North American countries

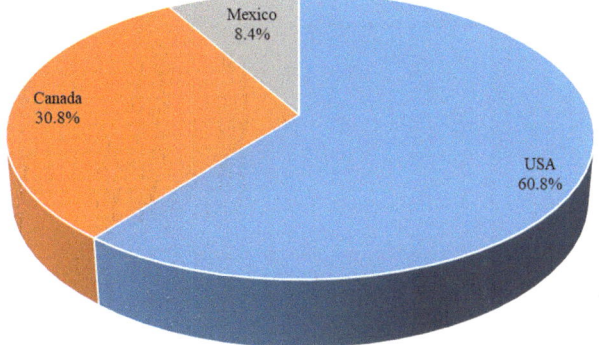

Fig. 4.12 Pie chart of undiscovered recoverable oil and gas resources in major North American countries

resources in Mexico are 35.9×10^8 t of oil equivalent, of which oil accounts for 62.2%, condensate accounts for 15.0%, and gas accounts for 22.8%.

4.1.3.2 Distribution of Undiscovered Oil and Gas Resources in Major Basins

The undiscovered oil and gas resources in North America are mainly distributed in the Gulf of Mexico Deepwater Basin, Alaska North Slope Basin, Gulf Coast Basin, and Scotia Basin. The Gulf of Mexico Deepwater Basin, Alaska North Slope Basin and Gulf Coast Basin rank among the top three basins in terms of the volumes of undiscovered oil and gas resources. The undiscovered oil and gas resources in these three basins account for 50.5% of the total undiscovered oil and gas resources in North America (Figs. 4.13 and 4.14).

The total undiscovered oil and gas resources in the Gulf of Mexico Deepwater Basin are 117.7×10^8 t of oil equivalent, of which oil accounts for 31.9%, condensate accounts for 24.2%, and gas accounts for 43.9%. The total undiscovered oil and gas resources in the Alaska North Slope Basin are 52.4×10^8 t of oil equivalent, of which oil accounts for 72.5%, condensate accounts for 7.7%, and gas accounts for 19.8%. The total undiscovered oil and gas resources in the Gulf Coast Basin are 45.1×10^8 t of oil equivalent, of which oil accounts for 4.5%, condensate accounts for 15.6%, and gas accounts for 79.9%.

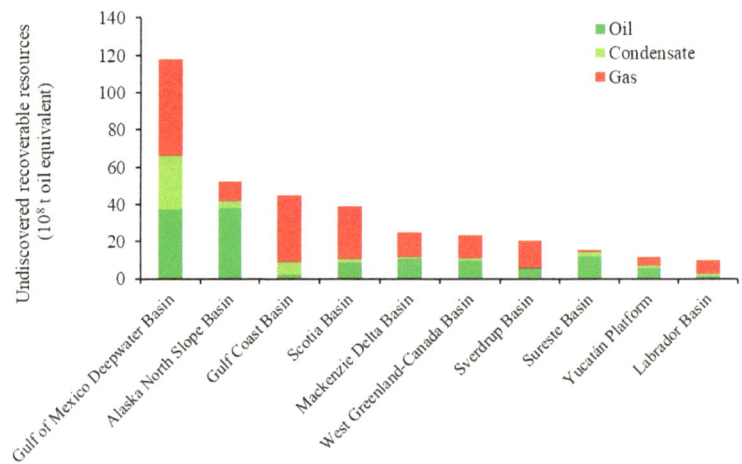

Fig. 4.13 Histogram of undiscovered recoverable oil and gas resources in major basins in North America

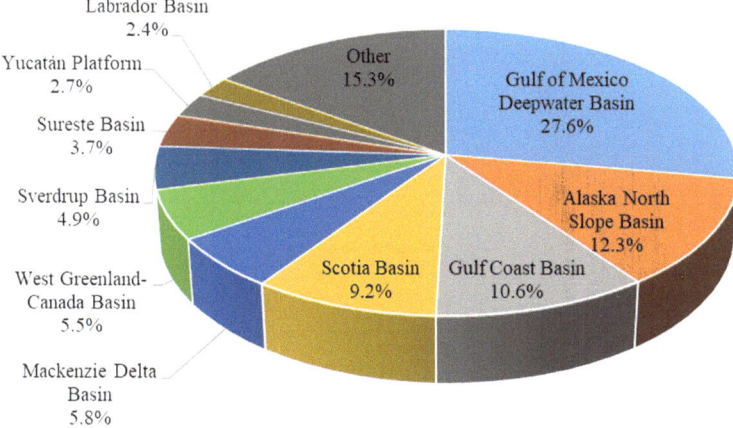

Fig. 4.14 Pie chart of undiscovered recoverable oil and gas resources in major basins in North America

4.1.3.3   Distribution of Undiscovered Onshore and Offshore Resources

In North America, the undiscovered offshore oil and gas resources are two times the undiscovered onshore oil and gas resources, and the undiscovered offshore and onshore oil and gas resources account for 66.4% and 33.6%, respectively. The volumes of undiscovered offshore and onshore oil resources are slightly smaller than those of undiscovered offshore and onshore gas resources (Fig. 4.15).

The undiscovered onshore oil and gas resources in North America are 143.2 × 10^8 t of oil equivalent, of which oil accounts for 39.9%, condensate accounts for

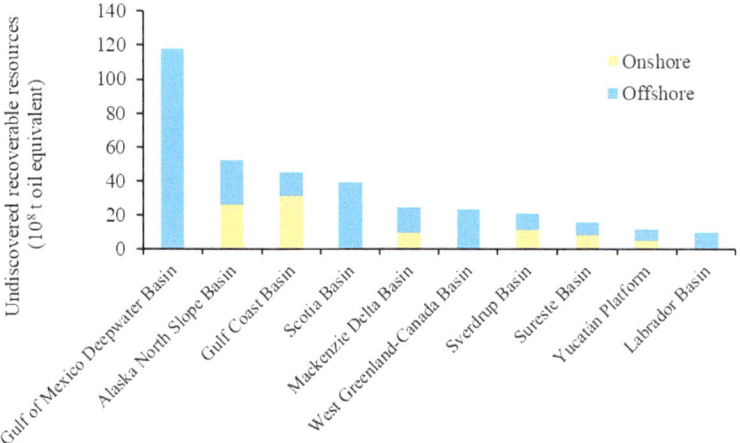

Fig. 4.15 Histogram of undiscovered recoverable oil and gas resources distributed in onshore and offshore areas in major basins in North America

10.6%, and gas accounts for 49.5%. The undiscovered offshore oil and gas resources are 283.0×10^8 t of oil equivalent, of which oil accounts for 36.4%, condensate accounts for 14.4%, and gas accounts for 49.2%.

4.2 Unconventional Oil and Gas Resources

The recoverable unconventional oil and gas resources in North America are 2440.1 $\times 10^8$ t of oil equivalent, accounting for 33.7% of the world's total recoverable unconventional resources and including seven types of resources, namely, oil shale, heavy oil, oil sands, shale oil, shale gas, tight gas, and CBM.

The recoverable unconventional oil resources in North America are 1586.2×10^8 t of oil equivalent, accounting for 37.4% of the world's total recoverable unconventional oil resources. These unconventional oil resources in North America include 544.5×10^8 t of oil shale, accounting for 34.3%, 403.4×10^8 t of oil sands, accounting for 25.4%, 324.7×10^8 t of heavy oil, accounting for 20.5%, and 313.6×10^8 t of shale oil, accounting for 19.8% (Figs. 4.16 and 4.17). Among the recoverable unconventional oil resources worldwide, 63.1% oil sands, 35.3% oil shale, 25.5% heavy oil, and 41.8% shale oil are concentrated in North America.

The recoverable unconventional gas resources in North America are 99.9×10^{12} m^3, accounting for 28.4% of the world's total unconventional gas resources. The recoverable shale gas resources are 74.3×10^{12} m^3, accounting for 74.4% of the total unconventional gas resources in North America; the recoverable CBM resources are 17.0×10^{12} m^3, accounting for 17.0%; and the recoverable tight gas resources are 8.6×10^{12} m^3, accounting for 8.6% (Figs. 4.18 and 4.19). Among the recoverable unconventional gas resources worldwide, 13.6% tight gas, 36.1% CBM and 31.0% shale gas are concentrated in North America.

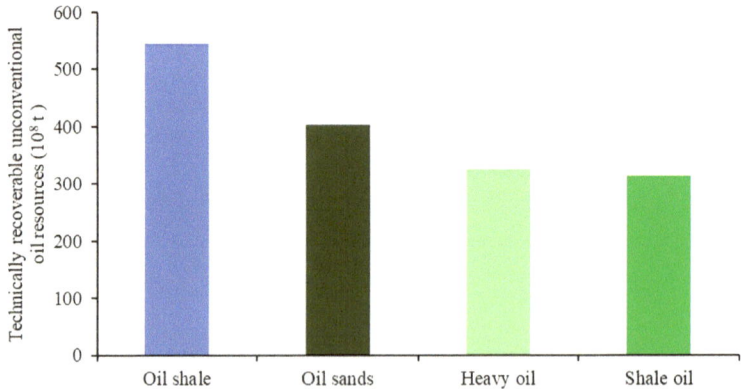

Fig. 4.16 Histogram of technically recoverable unconventional oil resources in North America

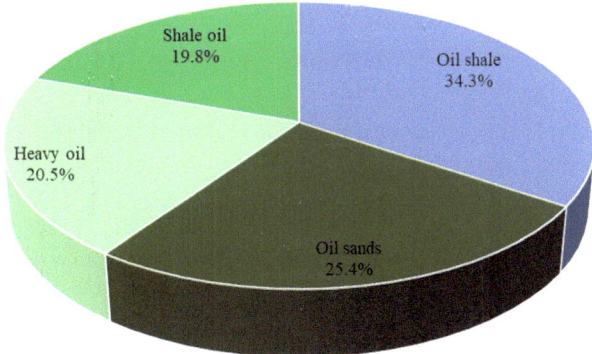

Fig. 4.17   Pie chart of technically recoverable unconventional oil resources in North America

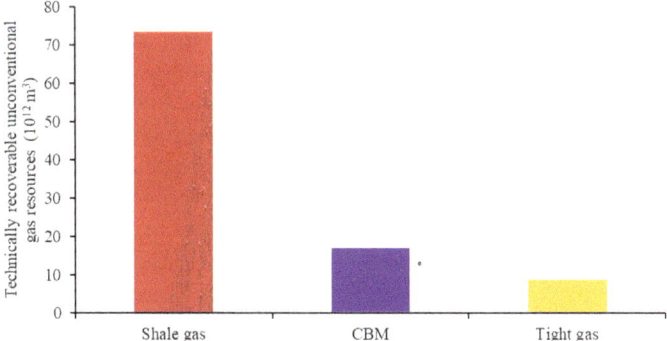

Fig. 4.18   Histogram of technically recoverable unconventional gas resources in North America

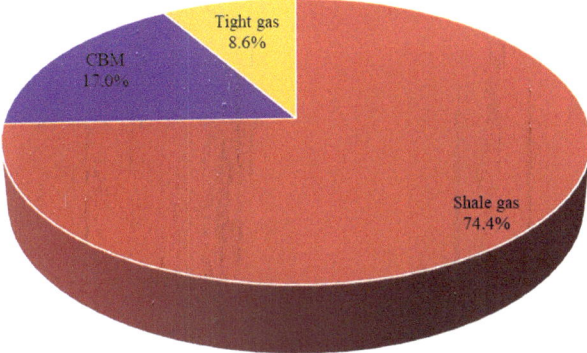

Fig. 4.19   Pie chart of technically recoverable unconventional gas resources in North America

4.2.1 *Distribution of Recoverable Unconventional Oil and Gas Resources in Major Countries (Regions)*

4.2.1.1 Distribution of Recoverable Unconventional Oil Resources in Major Countries (Regions)

The unconventional oil resources in North America are mainly distributed in the USA, Canada, and Mexico. The main type of unconventional oil resource in the USA is oil shale, the main unconventional oil resources in Canada are oil sands, and the main unconventional oil resources in Mexico are heavy oil resources. The unconventional oil resources in the USA are 1024.3×10^8 t, accounting for 64.6% of the unconventional oil resources in North America and 25.3% of the world's total unconventional oil resources. The main types of unconventional oil resources in the USA are oil shale, shale oil, and heavy oil. The recoverable oil shale are 536.3×10^8 t, the recoverable shale oil resources are 291.8×10^8 t, the recoverable heavy oil resources are 186.0×10^8 t, and the recoverable oil sands are 10.3×10^8 t. The recoverable unconventional oil resources in Canada amount to 413.9×10^8 t, accounting for 26.1% of the total unconventional oil resources in North America and including 8.5×10^8 t of oil shale, 12.3×10^8 t of shale oil, and 393.1×10^8 t of oil sands. The recoverable unconventional oil resources in Mexico are 148.2×10^8 t, accounting for 9.3% of the unconventional oil resources in North America and 3.7% of the world's total unconventional oil resources. Mexico's unconventional oil resources include heavy oil and shale oil. In Mexico, the recoverable heavy oil resources are 138.7×10^8 t, and the recoverable shale oil resources are 9.5×10^8 t (Figs. 4.20 and 4.21).

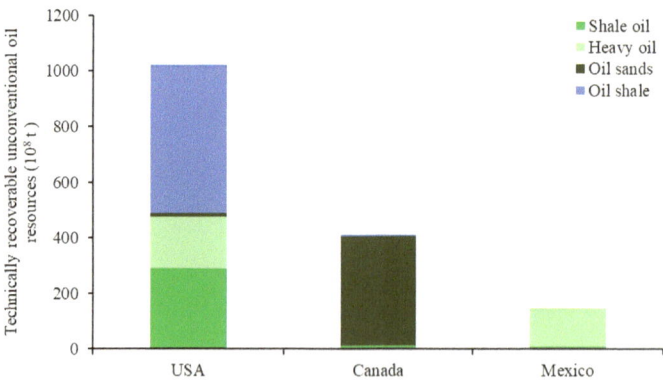

Fig. 4.20 Histogram of technically recoverable unconventional oil resources in major countries of North America

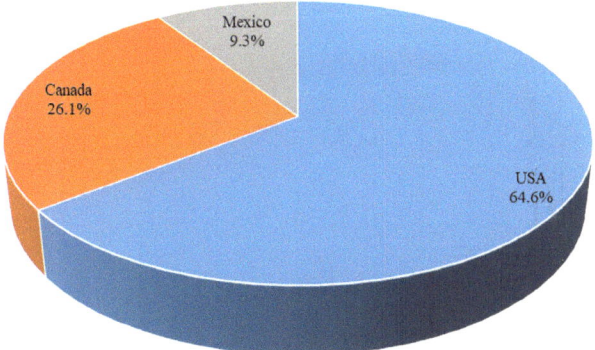

Fig. 4.21 Pie chart of technically recoverable unconventional oil resources in major countries of North America

4.2.1.2 Distribution of Recoverable Unconventional Gas Resources in Major Countries (Regions)

The unconventional gas resources in North America are mainly distributed in the USA and Canada. The recoverable unconventional gas resources in the USA are 71.5×10^{12} m^3, accounting for 71.4% of the unconventional gas resources in North America and 28.6% of the world's total unconventional gas resources. The recoverable resources of shale gas, CBM and tight gas in the USA are 58.4×10^{12} m^3, 7.7×10^{12} m^3 and 5.4×10^{12} m^3, respectively. The recoverable unconventional gas resources in Canada are 27.8×10^{12} m^3, accounting for 27.8% of the total unconventional gas resources in North America. The recoverable resources of shale gas, CBM and tight gas in Canada are 15.3×10^{12} m^3, 9.3×10^{12} m^3 and 3.1×10^{12} m^3, respectively. The recoverable unconventional gas resources in Mexico are 0.8×10^{12} m^3, all of which are shale gas, accounting for 0.8% of the total unconventional gas resources in North America (Figs. 4.22 and 4.23).

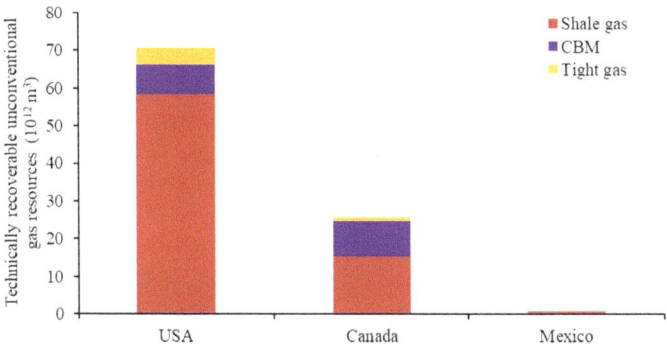

Fig. 4.22 Histogram of technically recoverable unconventional gas resources in major countries of North America

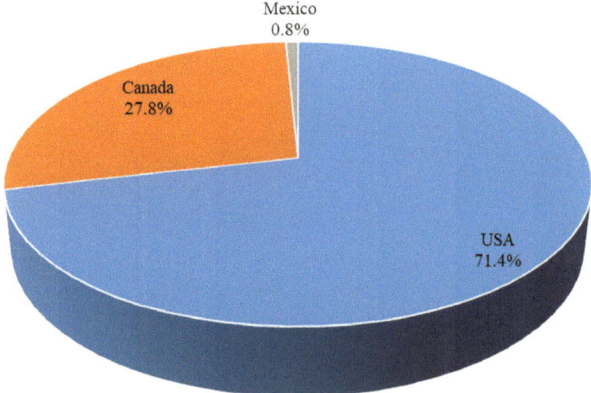

Fig. 4.23 Pie chart of technically recoverable unconventional gas resources in major countries of North America

4.2.2 Distribution of Recoverable Unconventional Oil and Gas Resources in Major Basins

In North America, 89.4% of unconventional oil resources and 93.6% of unconventional gas resources are concentrated in the top ten basins ranked according to the volumes of unconventional resources therein.

4.2.2.1 Distribution of Recoverable Unconventional Oil Resources in Major Basins

The recoverable unconventional oil resources in North America are mainly concentrated in 28 basins, of which the Alberta Basin in Canada ranks first in terms of recoverable unconventional oil resources. The recoverable unconventional oil resources in the Alberta Basin are 411.3×10^8 t, accounting for 25.9% of the total unconventional oil resources in North America. Among the recoverable unconventional oil resources in the Alberta Basin, the recoverable oil sands, which are the main unconventional oil resources, are 392.9×10^8 t, the recoverable shale oil resources are 10.1×10^8 t, and the recoverable oil shale is 8.3×10^8 t. The Uinta Basin ranks second, and the recoverable unconventional oil resources in this basin are 233.3×10^8 t, accounting for 14.7% of the total unconventional oil resources in North America. In the Uinta Basin, the recoverable oil shale, which are the main unconventional oil resources, are 230.6×10^8 t. The Piceance Basin ranks third, and the recoverable unconventional oil resources in this basin are 146.8×10^8 t, accounting for 9.3% of the total unconventional oil resources in North America. In the Piceance Basin, the recoverable oil shale, which are the main unconventional oil resources, are 137.6×10^8 t, and the recoverable shale oil resources are 9.2×10^8 t. The Permian Basin, from which shale oil has been successfully produced, ranks fifth in terms of total recoverable unconventional oil resources. The amount of recoverable shale oil in the Permian Basin reaches up to 107.5×10^8 t, which is the highest in North America (Figs. 4.24 and 4.25).

In North America, 80.2% of recoverable heavy oil resources are distributed in the Tampico-Misantla Basin, San Juan Basin, Yucatán Basin, and Ventura Basin; 97.4% of recoverable oil sands are distributed in the Alberta Basin; and 92.0% of recoverable shale oil resources are distributed in the Permian Basin, Gulf Coast Basin, Appalachian Basin, Williston Basin, Denver Basin, Alberta Basin, Piceance Basin, and Tampico-Misantla Basin.

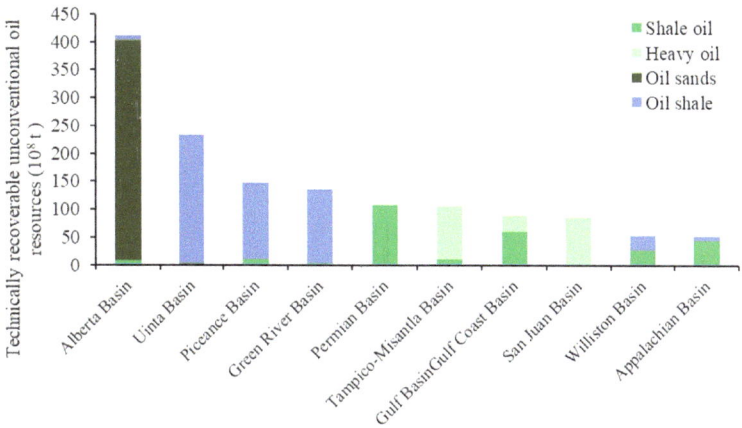

Fig. 4.24 Histogram of technically recoverable unconventional oil resources in major basins in North America

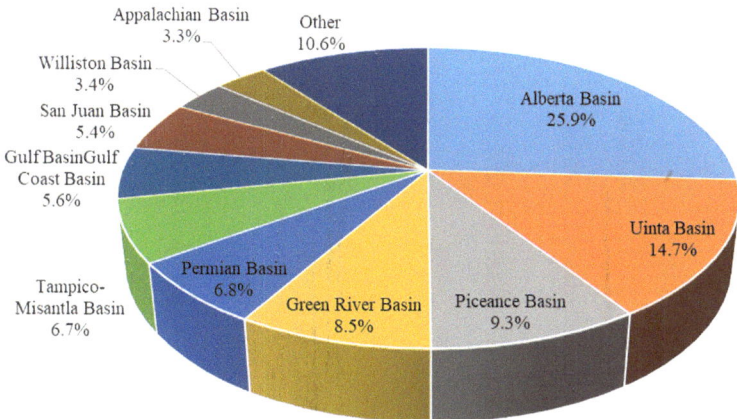

Fig. 4.25 Pie chart of technically recoverable unconventional oil resources in major basins in North America

4.2.2.2 Distribution of Recoverable Unconventional Gas Resources in Major Basins

The unconventional gas resources in North America are mainly distributed in 16 basins. The recoverable unconventional gas resources in the Alberta Basin are 27.6×10^{12} m³, accounting for 27.6% of the total recoverable unconventional gas resources in North America. The recoverable resources of shale gas, CBM and tight gas in the Alberta Basin are 15.2×10^{12} m³, 9.3×10^{12} m³ and 3.1×10^{12} m³, respectively. The Gulf Coast Basin in the USA ranks second in terms of unconventional gas resources. The recoverable unconventional gas resources in the Gulf Coast Basin are 25.0×10^{12} m³, accounting for 25.0% of the total recoverable unconventional gas resources in North America. The recoverable shale gas resources, which are the main type of unconventional gas resource in this basin, are 24.4×10^{12} m³. The Appalachian Basin in the USA ranks third in terms of unconventional gas resources. The recoverable unconventional gas resources in the Appalachian Basin are 19.9×10^{12} m³, accounting for 19.9% of the total recoverable unconventional gas resources in North America. The main type of unconventional gas resource in this basin is shale gas, and the recoverable shale gas resources in this basin are 17.4×10^{12} m³. Other recoverable unconventional gas resources are distributed in the Permian Basin, Piceance Basin, San Juan Basin, Williston Basin, Denver Basin, Green River Basin, Wyoming Thrust Belt, Uinta Basin, and other basins (Figs. 4.26 and 4.27).

In North America, 87.5% of recoverable shale gas resources are concentrated in the Gulf Coast Basin, Appalachian Basin, Alberta Basin, and Permian Basin; 81.4% of recoverable CBM resources are distributed in the Alberta Basin, Piceance Basin, Williston Basin, and San Juan Basin; and 72.2% of tight gas resources are distributed across the Appalachian Basin, Alberta Basin, San Juan Basin, and Green River Basin.

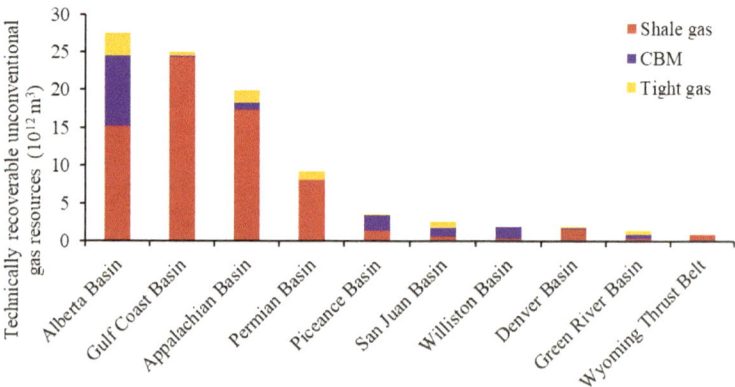

Fig. 4.26 Histogram of technically recoverable unconventional gas resources in major basins in North America

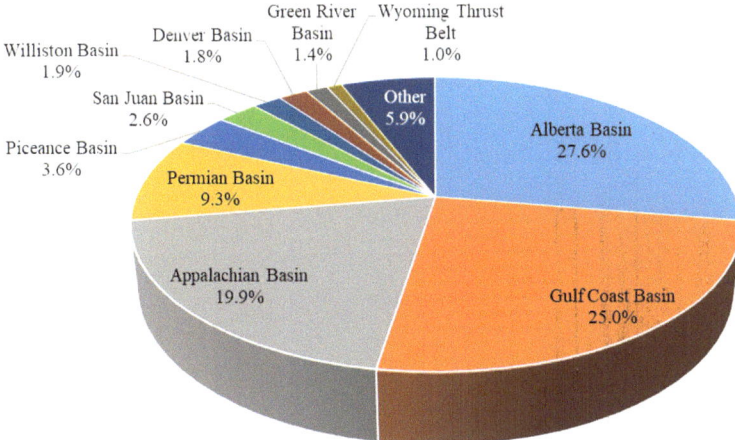

Fig. 4.27 Pie chart of technically recoverable unconventional gas resources in major basins in North America

Chapter 5
Distribution of Oil and Gas Resources in Central and South America

Central and South America refers to the region south of Mexico and north of Antarctica across the offshore, including 31 countries such as Venezuela, Guyana, and Brazil. The total area of Central and South America is 1820×10^4 km^2 (including nearby islands), accounting for 12.0% of the global land area. More than 100 sedimentary basins have developed in Central and South America, where foreland basins, passive margin basins, and cratonic basins are the main types of basins. Central and South America has abundant oil and gas resources accounting for 12.1% of the global oil and gas resources. The total recoverable oil and gas resources in Central and South America are 2258.6×10^8 t.

5.1 Conventional Oil and Gas Resources

The conventional oil and gas resources in Central and South America are 1164.3×10^8 t of oil equivalent, accounting for 10.2% of the world's total conventional resources. The recoverable conventional oil and gas resources in Central and South America are 499.8×10^8 t of oil equivalent, accounting for 7.7% of the world's total recoverable conventional resources. The cumulative oil and gas production in Central and South America is 246.1×10^8 t of oil equivalent, accounting for 9.4% of the global cumulative oil and gas production. The remaining recoverable reserves of oil and gas in Central and South America are 253.7×10^8 t of oil equivalent, accounting for 6.5% of the world's total remaining recoverable reserves. The future reserve growth from discovered oil and gas fields in Central and South America is 77.5×10^8 t of oil equivalent, accounting for 6.9% of the global reserve growth. The undiscovered recoverable oil and gas resources in Central and South America are 587.1×10^8 t of oil equivalent, accounting for 15.5% of the world's total undiscovered recoverable resources.

© The Author(s) 2024   93
L. Dou et al., *Global Oil and Gas Resources: Potential and Distribution*,
https://doi.org/10.1007/978-981-97-4756-6_5

5.1.1   *Distribution of Remaining Recoverable Reserves*

The total remaining recoverable reserves of oil and gas in Central and South America are 253.7×10^8 t of oil equivalent, including 153.3×10^8 t of oil, accounting for 60.4%, 10.6×10^8 t of condensate, accounting for 4.2%, and 10.5×10^{12} m³ of gas, accounting for 35.4%.

5.1.1.1   Distribution of Remaining Recoverable Reserves in Major Countries (Regions)

Most remaining recoverable reserves of oil and gas in Central and South America are distributed in Brazil and Venezuela, and the proportions of remaining recoverable reserves in other countries account are relatively small. The remaining recoverable reserves of oil and gas in Brazil are 91.5×10^8 t of oil equivalent, of which oil accounts for 71.8%, condensate accounts for 2.4%, and gas accounts for 25.8%. The remaining recoverable reserves of oil and gas in Venezuela are 86.2×10^8 t of oil equivalent, ranking second, of which oil accounts for 58.5%, condensate accounts for 5.0%, and gas accounts for 36.5%. The remaining recoverable oil and gas reserves in other countries are 76.0×10^8 t of oil equivalent, accounting for only 30% of the total remaining recoverable reserves in Central and South America (Figs. 5.1 and 5.2).

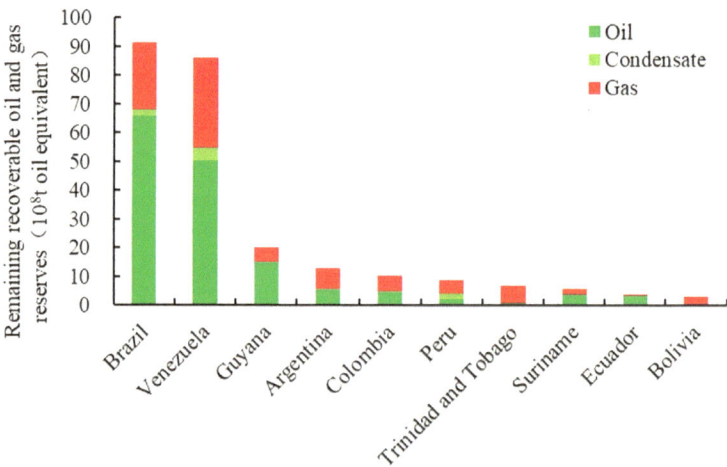

Fig. 5.1 Histogram of remaining recoverable reserves of oil and gas in major countries (regions) in Central and South America

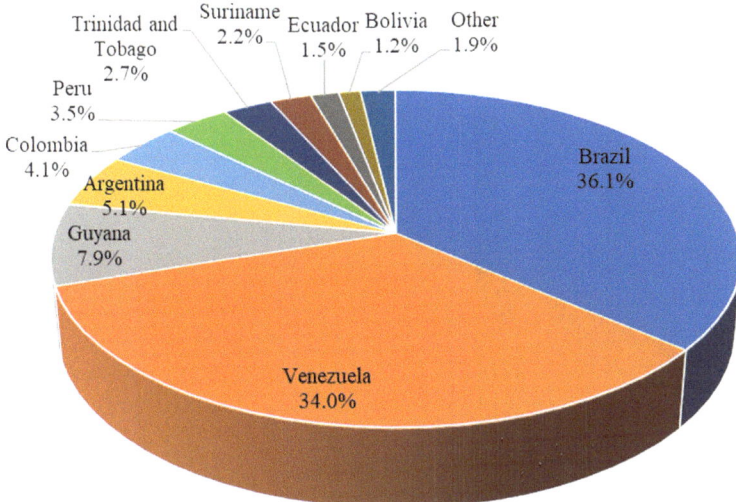

Fig. 5.2 Pie chart of remaining recoverable reserves of oil and gas in major countries (regions) in Central and South America

5.1.1.2 Distribution of Remaining Recoverable Reserves in Major Basins

The remaining recoverable reserves of oil and gas in Central and South America are mainly concentrated in the Santos Basin, Eastern Venezuela Basin, Maracaibo Basin, and Campos Basin. The remaining recoverable reserves of oil and gas in the Santos Basin account for 21.8%. The Eastern Venezuela Basin and Maracaibo Basin rank second and third in terms of remaining recoverable resources. The remaining recoverable reserves of oil and gas in these two basins account for 15.8% and 14.1%, respectively (Figs. 5.3 and 5.4).

The remaining recoverable reserves of oil and gas in the Santos Basin are 55.3 $\times 10^8$ t of oil equivalent, of which oil accounts for 77.3%, condensate accounts for 1.9%, and gas accounts for 20.8%. The remaining recoverable reserves of oil and gas in the Eastern Venezuela Basin are 40.1 $\times 10^8$ t of oil equivalent, of which oil accounts for 46.2%, condensate accounts for 8.5%, and gas accounts for 45.4%. The remaining recoverable reserves of oil and gas in the Maracaibo Basin are 35.8 $\times 10^8$ t of oil equivalent, of which oil accounts for 78.2%, condensate accounts for 1.5%, and gas accounts for 20.4%.

5.1.1.3 Distribution of Remaining Recoverable Reserves in Onshore and Offshore Areas

Most remaining recoverable reserves of oil and gas in Central and South America are distributed offshore. The remaining recoverable reserves of offshore oil and gas

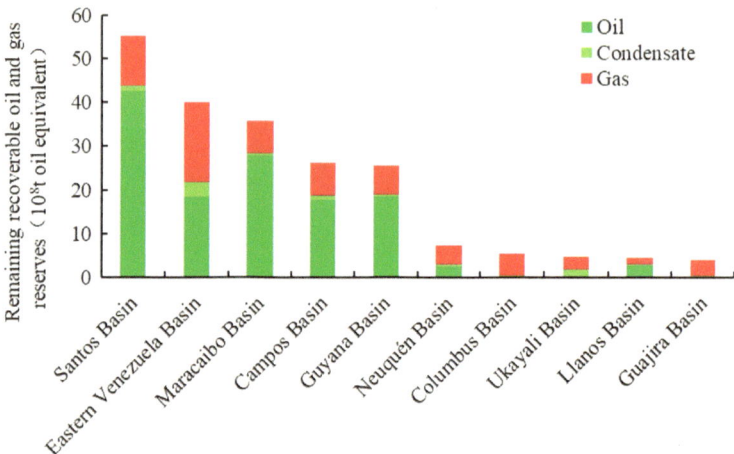

Fig. 5.3 Histogram of remaining recoverable reserves in major basins in Central and South America

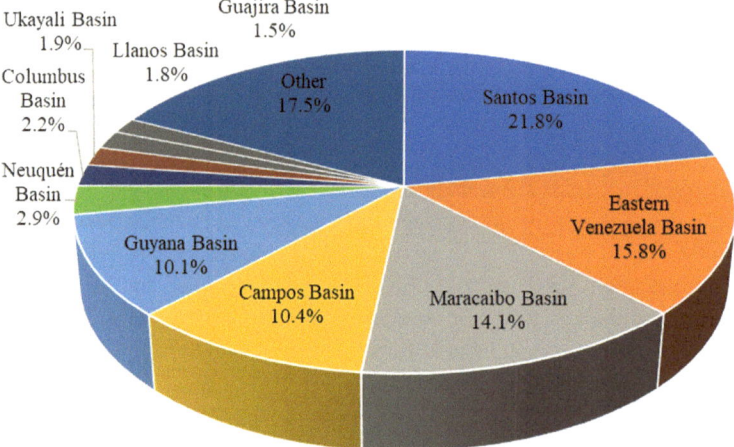

Fig. 5.4 Pie chart of remaining recoverable reserves in major basins in Central and South America

in Central and South America account for 65.0% of the total remaining recoverable reserves in this region. The remaining recoverable oil reserves in both onshore and offshore areas are more than the remaining recoverable gas reserves in such areas. The remaining recoverable reserves of onshore oil and gas are 88.8×10^8 t of oil equivalent, of which oil accounts for 86.6%, condensate accounts for 1.4%, and gas accounts for 12.0%. The remaining recoverable reserves of offshore oil and gas are 164.8×10^8 t of oil equivalent, of which oil accounts for 69.4%, condensate accounts for 1.8%, and gas accounts for 28.8%. (Fig. 5.5).

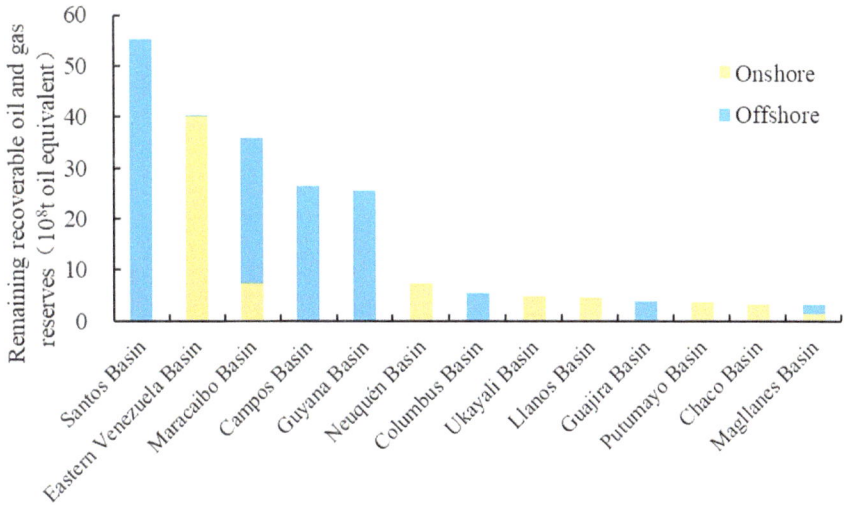

Fig. 5.5 Histogram of onshore and offshore remaining recoverable reserves in major basins in Central and South America

5.1.2 Trend of Reserve Growth from Discovered Oil and Gas Fields

The total future reserve growth from discovered oil and gas fields in Central and South America are 77.5×10^8 t of oil equivalent, including 51.1×10^8 t of oil, accounting for 65.9%, 2.6×10^8 t of condensate, accounting for 3.4%, and 2.8×10^{12} m^3 of gas, accounting for 30.7%.

5.1.2.1 Distribution of Reserve Growth in Major Countries (Regions)

The future reserve growth in discovered oil and gas fields are mainly distributed in Brazil and Venezuela. The reserve growth of oil and gas in these two countries account for 45.1% and 33.8% of the total reserve growth in Central and South America, respectively. The discovered oil and gas fields in Bolivia, Peru and Colombia have certain potential for reserve growth, while those in other countries have small potential for reserve growth (Figs. 5.6 and 5.7).

The future reserve growth from discovered oil and gas fields in Brazil are 34.9 $\times 10^8$ t of oil equivalent, of which oil accounts for 72.9%, condensate accounts for 2.3%, and gas accounts for 24.8%. The future reserve growth from discovered oil and gas fields in Venezuela are 26.2×10^8 t of oil equivalent, of which oil accounts for 76.6%, condensate accounts for 1.6%, and gas accounts for 21.8%. The future reserve growth from discovered oil and gas fields in Bolivia are 8.2×10^8 t of oil

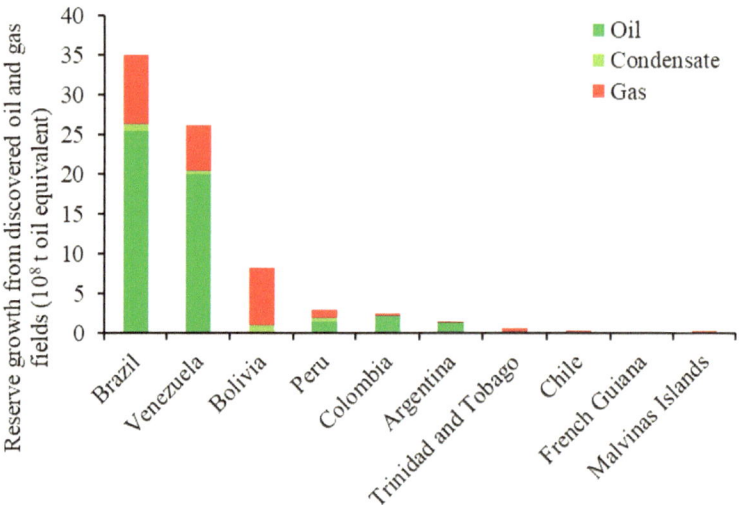

Fig. 5.6 Histogram of reserve growth in discovered oil and gas fields in major Central and South American countries (regions)

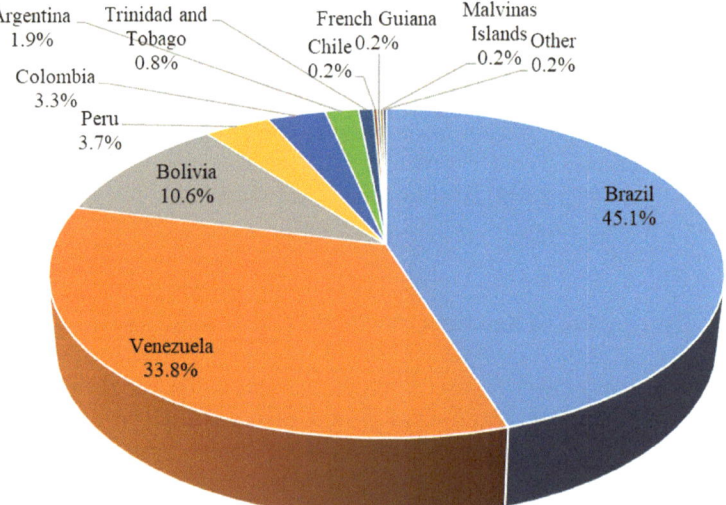

Fig. 5.7 Pie chart of reserve growth in discovered oil and gas fields in major Central and South American countries (regions)

equivalent, of which oil accounts for 1.7%, condensate accounts for 10.6%, and gas accounts for 87.7%.

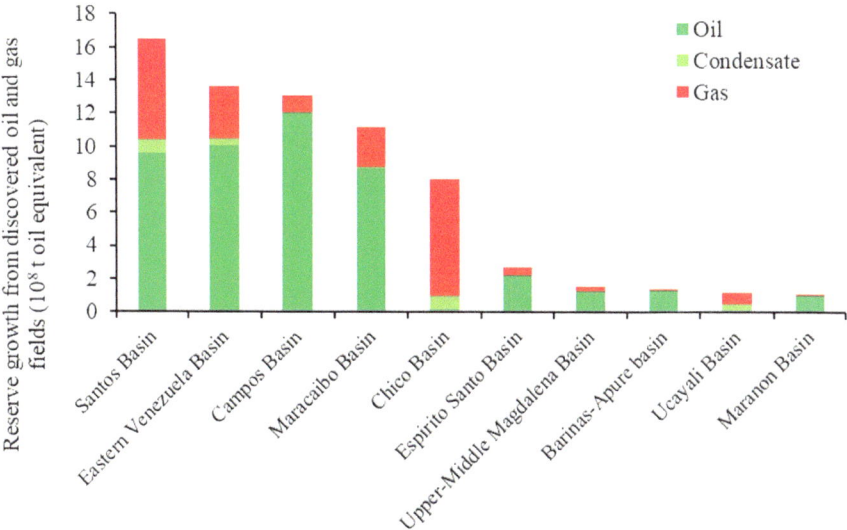

Fig. 5.8 Histogram of reserve growth in discovered oil and gas fields in major basins in Central and South America

5.1.2.2 Distribution of Reserve Growth in Major Basins

The future reserve growth in Central and South America mainly come from discovered oil and gas fields in the Santos Basin, Eastern Venezuela Basin, Campos Basin, and Maracaibo Basin. The Santos Basin, Eastern Venezuela Basin and Campos Basin are the top three basins in terms of future reserve growth. The reserve growth in discovered oil and gas fields in these three basins account for 55.6% of the total reserve growth in Central and South America (Figs. 5.8 and 5.9).

The potential of the Santos Basin for reserve growth in the future is 16.5×10^8 t of oil equivalent, of which oil accounts for 58.2%, condensate accounts for 4.8%, and gas accounts for 37.0%. The potential of the Eastern Venezuela Basin for reserve growth in the future is 13.6×10^8 t of oil equivalent, of which oil accounts for 73.6%, condensate accounts for 3.0%, and gas accounts for 23.4%. The potential of the Campos Basin for reserve growth in the future is 13.0×10^8 t of oil equivalent, of which oil accounts for 92.4% and gas accounts for 7.6%.

5.1.2.3 Distribution of Reserve Growth in Discovered Onshore and Offshore Fields

The reserve growth in discovered onshore and offshore fields in Central and South America account for 51.7% and 48.3% of the total reserve growth in this region, respectively (Fig. 5.10). The reserve growth in both onshore and offshore fields in

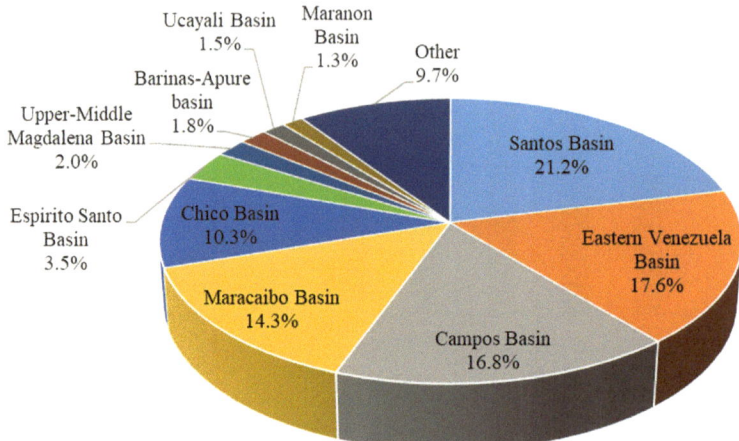

Fig. 5.9 Pie chart of reserve growth in discovered oil and gas fields in major basins in Central and South America

this region mainly consist of oil. The onshore and offshore additional oil reserves are about two times the onshore and offshore additional gas reserves.

The reserve growth from discovered onshore oil and gas fields in Central and South America is 40×10^8 t of oil equivalent, of which oil accounts for 72.3%, condensate accounts for 2.4%, and gas accounts for 25.3%. The reserve growth from discovered offshore oil and gas fields are 37.5×10^8 t of oil equivalent, of which oil accounts for 65.9%, condensate accounts for 3.4%, and gas accounts for 30.7%.

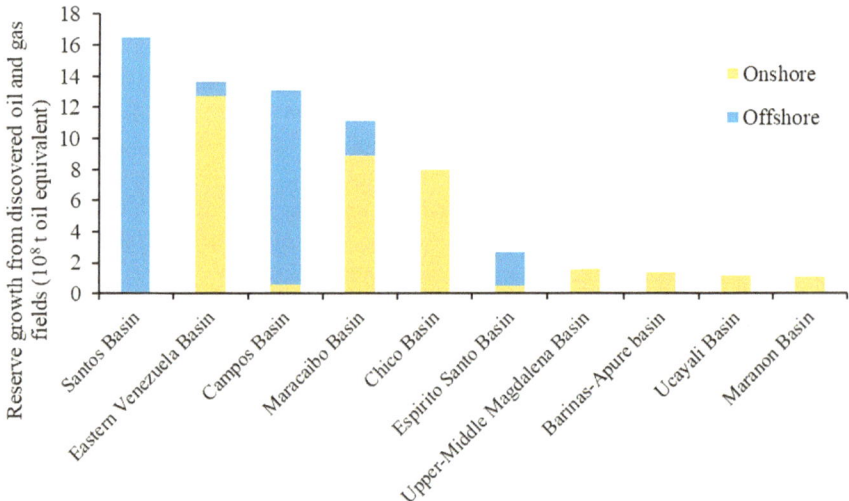

Fig. 5.10 Histogram of reserve growth in discovered onshore and offshore fields in major basins across Central and South America

5.1.3 Characteristics of Distributions of Undiscovered Oil and Gas Resources

The undiscovered oil and gas resources in Central and South America are 587.1 × 10^8 t of oil equivalent, including 435.5 × 10^8 t of oil, accounting for 74.2%, 20.6 × 10^8 t of condensate, accounting for 3.5%, and 15.3 × 10^{12} m^3 of gas, accounting for 22.3%.

5.1.3.1 Distribution of Undiscovered Oil and Gas Resources in Major Countries (Regions)

The undiscovered oil and gas resources in Central and South America are mainly distributed in Brazil, Argentina, Venezuela, Trinidad and Tobago, Bolivia, Guyana, and the Malvinas Islands, among which Brazil has the largest amounts of undiscovered oil and gas resources accounting for 59.1% of the total undiscovered oil and gas resources in this region (Figs. 5.11 and 5.12).

The undiscovered recoverable oil and gas resources in Brazil are 346.8 × 10^8 t of oil equivalent, of which oil accounts for 82.0%, condensate accounts for 1.3%, and gas accounts for 16.7%. The undiscovered recoverable oil and gas resources in Argentina

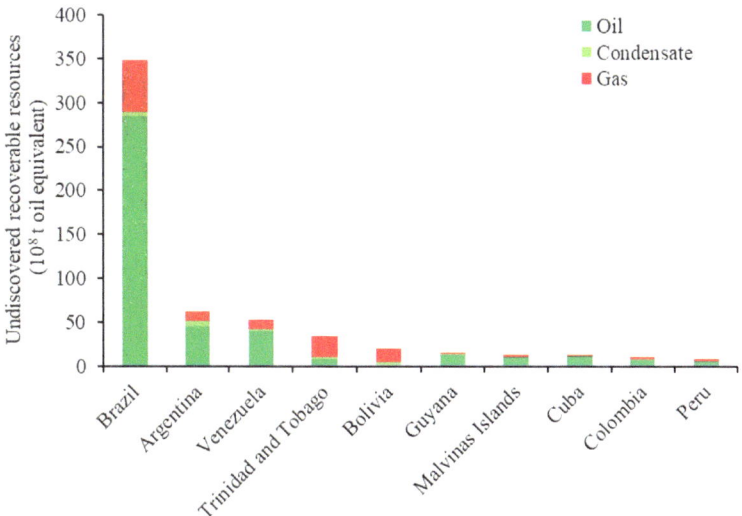

Fig. 5.11 Histogram of undiscovered recoverable oil and gas resources in major countries (regions) in Central and South America

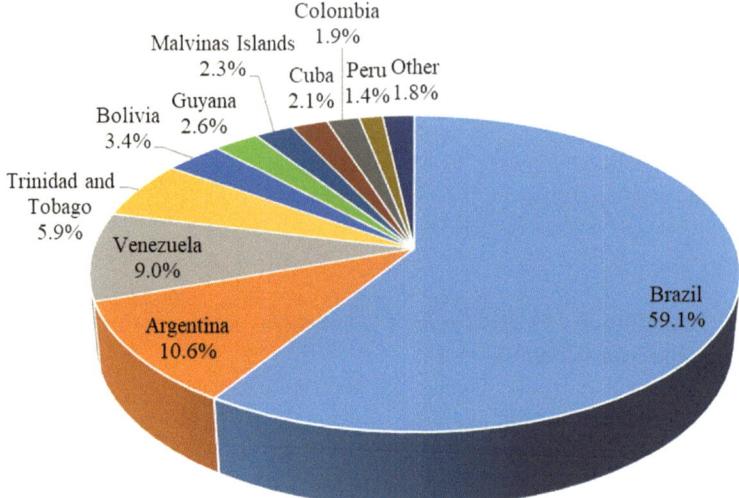

Fig. 5.12 Pie Chart of undiscovered recoverable oil and gas resources in major countries (regions) in Central and South America

are 62.2×10^8 t of oil equivalent, of which oil accounts for 72.8%, condensate accounts for 9.8%, and gas accounts for 17.4%. The undiscovered recoverable oil and gas resources in Venezuela are 52.9×10^8 t of oil equivalent, of which oil accounts for 76.3%, condensate accounts for 3.7%, and gas accounts for 20%.

5.1.3.2 Distribution of Undiscovered Oil and Gas Resources in Major Basins

In Central and South America, the Santos Basin and Campos Basin are two basins with the most undiscovered oil and gas resources. The undiscovered oil and gas resources in these two basins account for 51.3% of the total undiscovered oil and gas resources in Central and South America. The undiscovered recoverable oil and gas resources in the Santos Basin are 153.3×10^8 t of oil equivalent, of which oil accounts for 86.5% and gas accounts for 13.5%. The undiscovered recoverable oil and gas resources in the Campos Basin are 148.0×10^8 t of oil equivalent, of which oil accounts for 74.0%, condensate accounts for 0.7%, and gas accounts for 25.3%. The undiscovered recoverable oil and gas resources in the Eastern Venezuela Basin are 31.0×10^8 t of oil equivalent, of which oil accounts for 84.5%, condensate accounts for 2.4%, and gas accounts for 13.1% (Figs. 5.13 and 5.14).

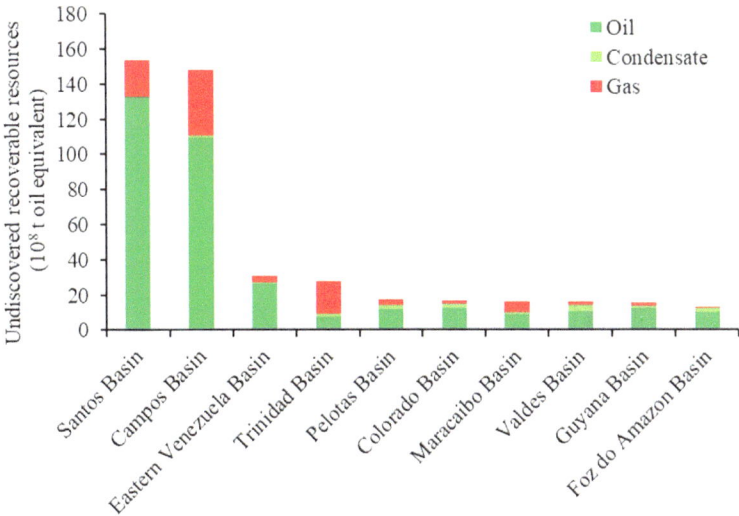

Fig. 5.13 Histogram of undiscovered recoverable oil and gas resources in major oil and gas-bearing basins in Central and South America

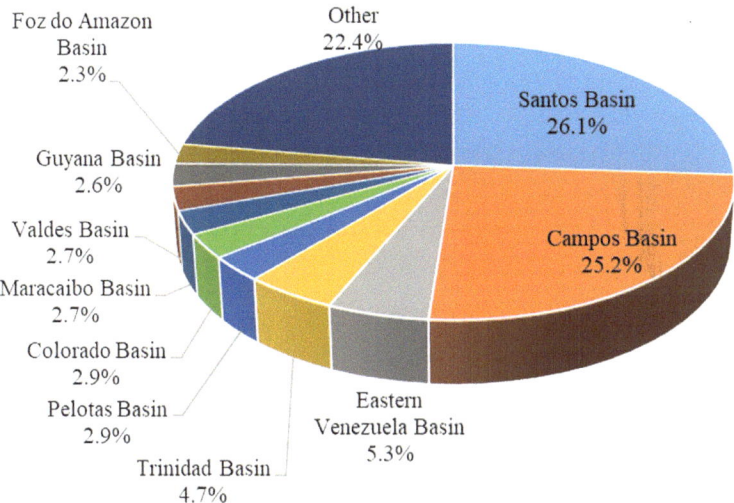

Fig. 5.14 Pie chart of undiscovered recoverable oil and gas resources in major oil and gas-bearing basins in Central and South America

5.1.3.3 Distribution of Undiscovered Oil and Gas Resources in Onshore and Offshore Areas

The undiscovered recoverable oil and gas resources in onshore areas in Central and South America are 151.2×10^8 t of oil equivalent (Fig. 5.15), of which oil accounts for

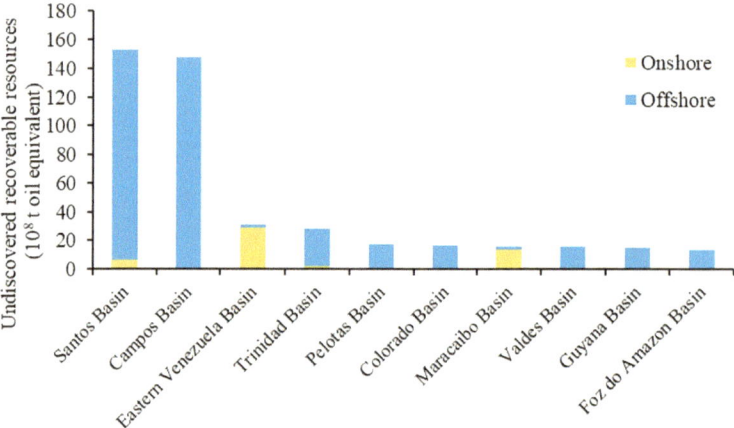

Fig. 5.15 Histogram of undiscovered recoverable oil and gas resources in onshore and offshore areas of major basins in Central and South America

64.4%, condensate accounts for 5.1%, and gas accounts for 30.5%. The undiscovered recoverable oil and gas resources in offshore areas are 435.9×10^8 t of oil equivalent, of which oil accounts for 76.5%, condensate accounts for 1.4%, and gas accounts for 22.1%.

5.2   Unconventional Oil and Gas Resources

The unconventional oil and gas resources in Central and South America include heavy oil, shale oil, oil shale, shale gas, CBM, and tight gas. The technically recoverable unconventional oil and gas resources in this region are $1,094.2 \times 10^8$ t of oil equivalent, accounting for 15.1% of the world's total recoverable unconventional resources. The technically recoverable unconventional oil resources in this region are 660.8×10^8 t, accounting for 15.6% of the world's total recoverable unconventional oil resources. In Central and South America, the recoverable heavy oil resources are 418.2×10^8 t, accounting for 32.8% of the global recoverable heavy oil resources; the recoverable oil shale is 153.2×10^8 t, accounting for 9.9% of the global recoverable oil shale; and the recoverable shale oil resources are 89.4×10^8 t, accounting for 11.9% of the global recoverable shale oil resources (Figs. 5.16 and 5.17).

The technically recoverable unconventional gas resources in Central and South America are 50.7×10^{12} m^3, accounting for 14.5% of the world's total recoverable unconventional gas resources. In this region, the technically recoverable shale gas resources are 40.5×10^{12} m^3, accounting for 16.9% of the world's total recoverable shale gas resources; the technically recoverable CBM resources are 0.03×10^{12} m^3,

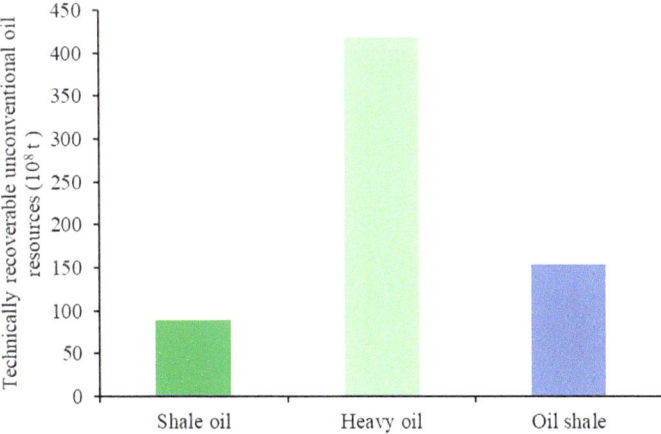

Fig. 5.16 Histogram of different types of technically recoverable unconventional oil resources in Central and South America

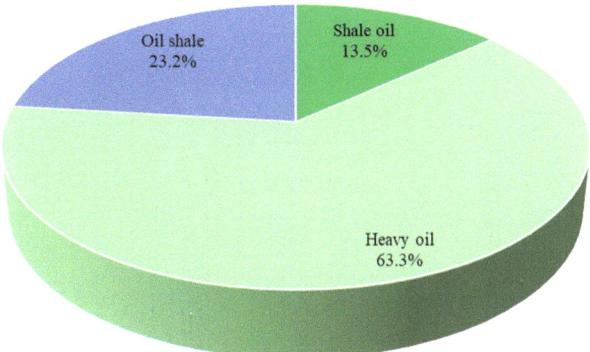

Fig. 5.17 Pie chart of different types of technically recoverable unconventional oil resources in Central and South America

accounting for 0.1% of the world's total recoverable CBM resources; and the technically recoverable tight gas resources are 10.2×10^{12} m^3, accounting for 16.2% of the world's total recoverable tight gas resources (Figs. 5.18 and 5.19).

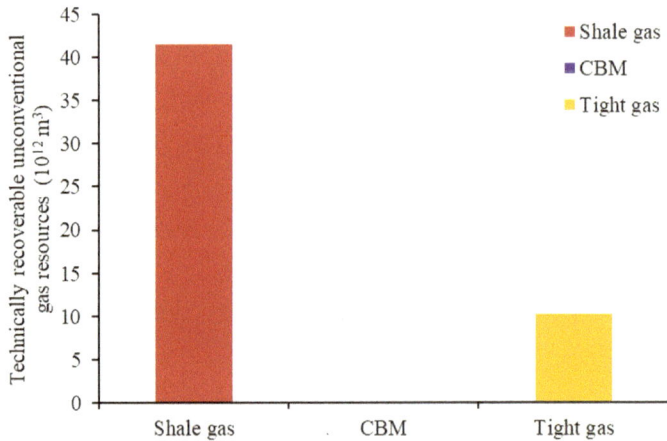

Fig. 5.18 Histogram of different types of technically recoverable unconventional gas resources in Central and South America

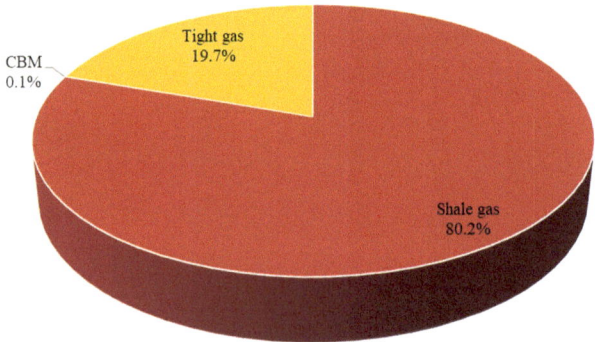

Fig. 5.19 Pie chart of different types of technically recoverable unconventional gas resources in Central and South America

5.2.1   Distribution of Recoverable Unconventional Oil and Gas Resources in Major Countries (Regions)

5.2.1.1   Distribution of Recoverable Unconventional Oil Resources in Major Countries (Regions)

The unconventional oil resources in Central and South America are mainly distributed in Venezuela, Brazil, Argentina, Colombia, Bolivia, Chile, and Peru (Figs. 5.20 and 5.21). The recoverable unconventional oil resources in Venezuela are 371.0×10^8 t, accounting for 56.1% of the total recoverable unconventional oil resources in Central and South America and including 350.2×10^8 t of heavy oil and 20.8×10^8 t of shale oil. The recoverable unconventional oil resources in Brazil are 207.7×10^8 t,

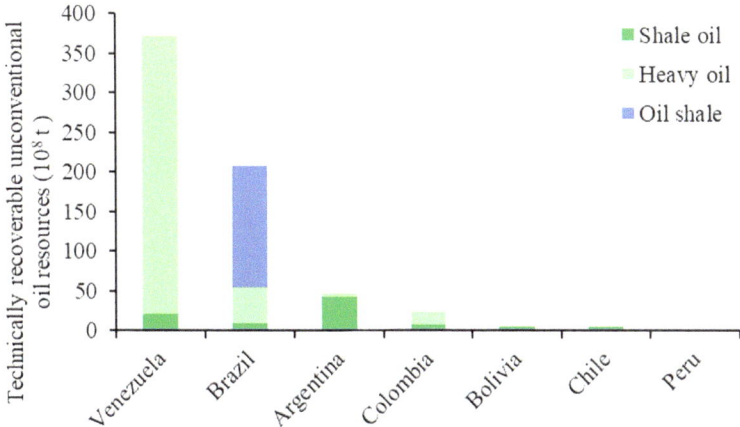

Fig. 5.20 Histogram of technically recoverable unconventional oil resources in major countries (regions) across Central and South America

accounting for 31.4% of the total recoverable unconventional oil resources in Central and South America and including 45.7×10^8 t of heavy oil, 8.8×10^8 t of shale oil, and 153.2×10^8 t of oil shale. The recoverable unconventional oil resources in Argentina are 47.0×10^8 t, accounting for 7.1% of the total recoverable unconventional oil resources in Central and South America and including 4.5×10^8 t of heavy oil and 42.5×10^8 t of shale oil. The recoverable unconventional oil resources in Colombia are 23.3×10^8 t, accounting for 3.5% of the total recoverable unconventional oil resources in Central and South America and including 15.8×10^8 t of heavy oil and 7.5×10^8 t of shale oil. The only recoverable unconventional oil resources in Bolivia and Chile are shale oil resources. The recoverable shale oil resources in these two countries are 5.3×10^8 t and 4.6×10^8 t, respectively. Heavy oil is the only recoverable unconventional oil resource in Peru. The recoverable heavy oil resources in Peru are 2.0×10^8 t.

5.2.1.2 Distribution of Recoverable Unconventional Gas Resources in Major Countries (Regions)

The unconventional gas resources in Central and South America are mainly distributed in Argentina, Venezuela, Brazil, Bolivia, Chile, and Colombia (Figs. 5.22 and 5.23). The recoverable unconventional gas resources in Argentina are 27.1×10^{12} m^3, accounting for 53.4% of the total recoverable unconventional gas resources in Central and South America and including 25.0×10^{12} m^3 of shale gas, and 2.1×10^{12} m^3 of tight gas. The recoverable unconventional gas resources in Venezuela are 7.7×10^{12} m^3, accounting for 15.1% of the total recoverable unconventional gas resources in Central and South America and including 5.8×10^{12} m^3 of shale gas, and 1.8×10^{12} m^3 of tight gas. The recoverable unconventional gas resources in Brazil

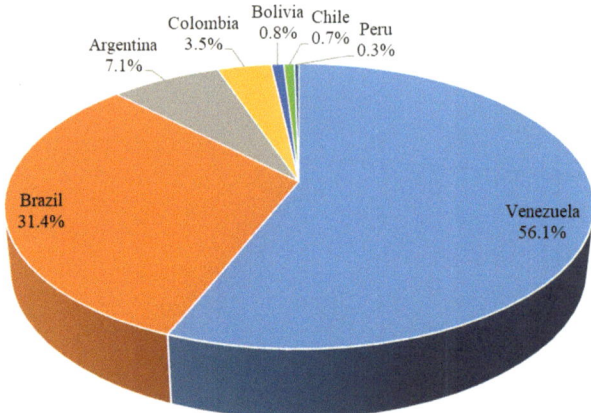

Fig. 5.21 Pie chart of technically recoverable unconventional oil resources in major countries (regions) across Central and South America

are 5.9×10^{12} m^3, accounting for 11.7% of the total recoverable unconventional gas resources in Central and South America and including 5.1×10^{12} m^3 of shale gas and 0.8×10^{12} m^3 of tight gas. The recoverable unconventional gas resources in Bolivia are 5.1×10^{12} m^3, accounting for 10.1% of the total recoverable unconventional gas resources in Central and South America and including 1.9×10^{12} m^3 of shale gas and 3.1×10^{12} m^3 of tight gas.

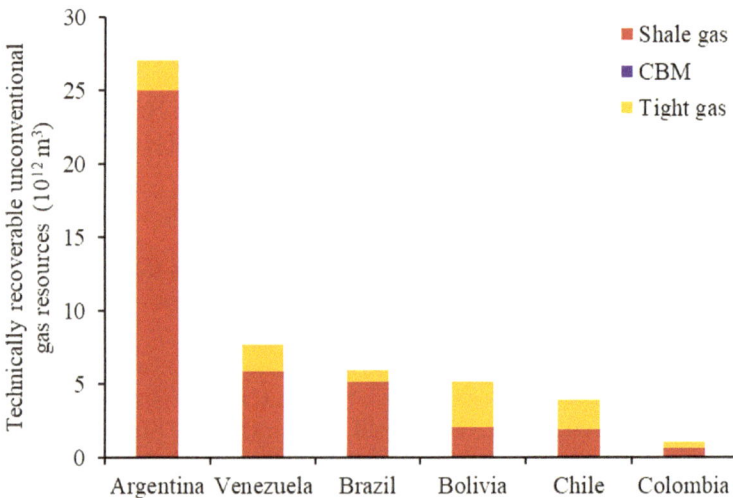

Fig. 5.22 Histogram of technically recoverable unconventional gas resources in major countries (regions) across Central and South America

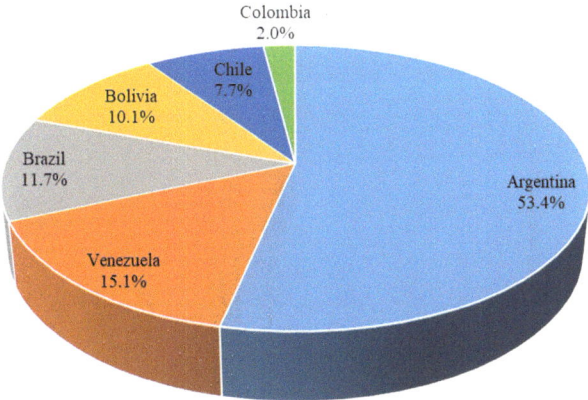

Fig. 5.23 Pie chart of technically recoverable unconventional gas resources in major countries (regions) across Central and South America

5.2.2 Distribution of Unconventional Oil and Gas Resources in Major Basins

The unconventional oil and gas resources in Central and South America are mainly distributed in 20 basins. Ten basins have unconventional oil and gas resources, while the other 10 basins only have unconventional oil resources. The Eastern Venezuela Basin has the largest amounts of unconventional oil and gas resources, which are 266.5×10^8 t, accounting for 24.4% of the total unconventional resources in Central and South America.

5.2.2.1 Distribution of Unconventional Oil Resources in Major Basins

The unconventional oil resources in Central and South America are mainly distributed in 20 basins. The Eastern Venezuela Basin has the largest amounts of unconventional oil resources, which are all heavy oil resources and account for 40.3% of the total unconventional oil resources in Central and South America (Figs. 5.24 and 5.25). The recoverable unconventional oil resources in the Amazon Basin rank second, reaching 114.4×10^8 t, which account for 17.3% of the total unconventional oil resources in Central and South America and include 113.4×10^8 t of oil shale and 1×10^8 t of shale oil. The recoverable unconventional oil resources in the Maracaibo Basin rank third, reaching 103.0×10^8 t, which account for 15.6% of the total unconventional oil resources in Central and South America and include 20.8×10^8 t of shale oil and 82.2×10^8 t of heavy oil. The recoverable unconventional oil resources in each of the following basins are more than 10×10^8 t: Eastern Venezuela Basin, Amazon Basin, Maracaibo Basin, Paraná Basin, Campos Basin, Neuquén Basin, Magallanes Basin, and Upper-Middle Magdalena Basin.

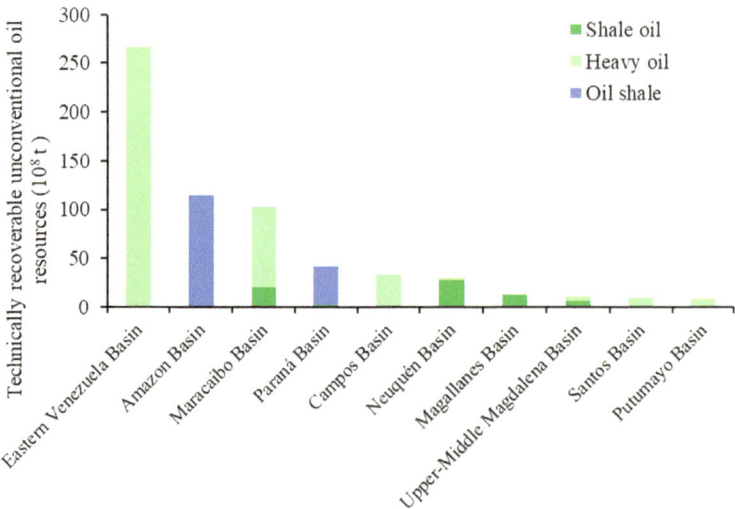

Fig. 5.24 Histogram of technically recoverable unconventional oil resources in major basins in Central and South America

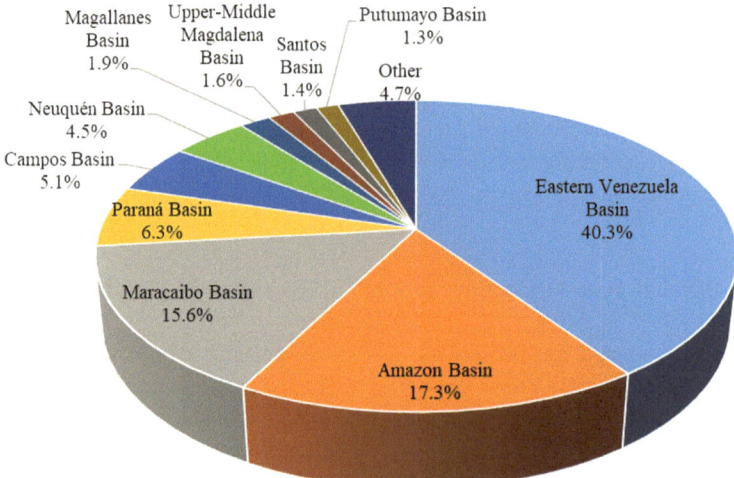

Fig. 5.25 Pie chart of technically recoverable unconventional oil resources in major basins in Central and South America

5.2.2.2   Distribution of Unconventional Gas Resources in Major Basins

The recoverable unconventional gas resources in Central and South America are mainly distributed in 10 basins, namely, the Neuquén Basin, Maracaibo Basin, Magallanes Basin, Amazon Basin, Chaco Basin, San Georges Basin, Chaco-Paraná Basin,

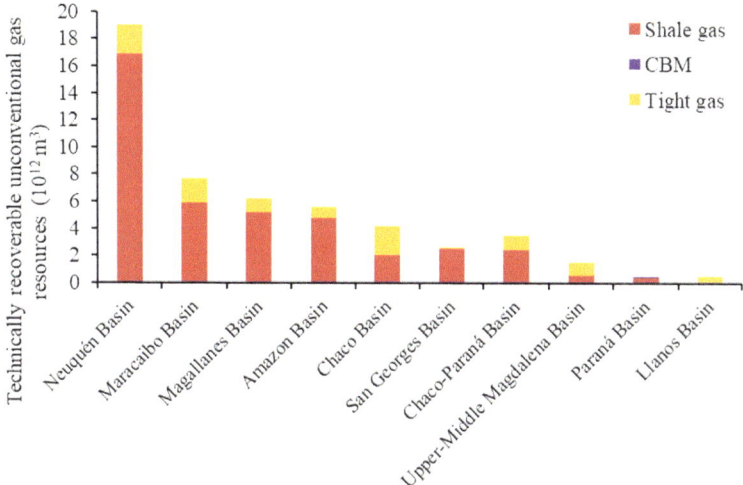

Fig. 5.26 Histogram of technically recoverable unconventional gas resources in major basins in Central and South America

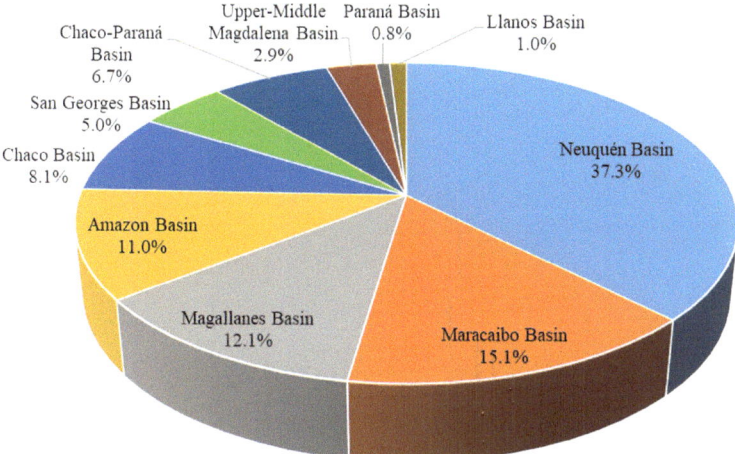

Fig. 5.27 Pie chart of technically recoverable unconventional gas resources in major basins in Central and South America

Upper-Middle Magdalena Basin, Paraná Basin, and Llanos Basin (Figs. 5.26 and 5.27). The recoverable unconventional gas resources in the Neuquén Basin are 18.9×10^{12} m^3, accounting for 37.3% of the total recoverable unconventional gas resources in Central and South America and including 16.9×10^{12} m^3 of shale gas and 2.0×10^{12} m^3 of tight gas. The recoverable unconventional gas resources in the Maracaibo Basin are 7.6×10^{12} m^3, accounting for 15.1% of the total recoverable unconventional gas resources in Central and South America and including 5.8×10^{12} m^3

of shale gas, and 1.8×10^{12} m^3 of tight gas. The recoverable unconventional gas resources in the Magallanes Basin are 6.2×10^{12} m^3, which account for 12.1% of the total recoverable unconventional gas resources in Central and South America, are mainly shale gas, and include 1.0×10^{12} m^3 of tight gas.

Chapter 6
Distribution of Oil and Gas Resources in Europe

Europe includes 46 countries and regions, such as the United Kingdom (UK), France, Germany and Norway, cover an area of $1016 \times 10^4 \, \text{km}^2$, and is the sixth largest continent worldwide. Forty-four sedimentary basins have developed in Europe. Foreland basins and cratonic basins are the main types of onshore basins, and rift basins and passive margin basins are the main types of offshore basins. 5.6% of the world's oil and gas resources are concentrated in Europe, and the recoverable oil and gas resources in Europe are 1051.9×10^8 t of oil equivalent.

6.1 Conventional Oil and Gas Resources

The recoverable conventional oil and gas resources in Europe are 534.3×10^8 t of oil equivalent, accounting for 4.7% of the world's total recoverable conventional resources. The recoverable reserves in Europe are 350.1×10^8 t, accounting for 5.4% of the world' total recoverable reserves. The remaining recoverable oil and gas reserves in Europe are 104.1×10^8 t of oil equivalent, accounting for 2.7% of the world's total remaining recoverable reserves. The estimated reserve growth from discovered oil and gas fields in the future is 32.3×10^8 t of oil equivalent, accounting for 2.9% of the world's total future reserve growth. The undiscovered recoverable oil and gas resources are 151.9×10^8 t of oil equivalent, accounting for 4.0% of the world's total undiscovered recoverable resources.

© The Author(s) 2024   113
L. Dou et al., *Global Oil and Gas Resources: Potential and Distribution*,
https://doi.org/10.1007/978-981-97-4756-6_6

6.1.1   Distribution of Remaining Recoverable Reserves

The remaining recoverable reserves of oil and gas in Europe are 104.1×10^8 t of oil equivalent, including 38.4×10^8 t of oil, 5.3×10^8 t of condensate, and 7.1×10^{12} m^3 of gas.

6.1.1.1   Distribution of Remaining Recoverable Reserves in Major Countries (Regions)

Norway has the largest amounts of remaining recoverable oil and gas reserves, reaching 31.8×10^8 t of oil equivalent, of which oil accounts for 41.8%, condensate accounts for 4.8%, and gas accounts for 53.4%. UK ranks second in terms of remaining recoverable oil and gas reserves. The remaining recoverable oil and gas reserves in UK are 18.6×10^8 t of oil equivalent, of which oil accounts for 51.6%, condensate accounts for 7.2%, and gas accounts for 41.2% (Figs. 6.1 and 6.2).

6.1.1.2   Distribution of Remaining Recoverable Reserves in Major Basin

The remaining recoverable reserves of oil and gas in Europe are mainly concentrated in the North Sea Basin, Dnieper-Donets Basin, Faroe-Shetland Basin, Northwest German Basin, Vøring Basin, South Carpathian Basin, Barents Sea Platform Basin,

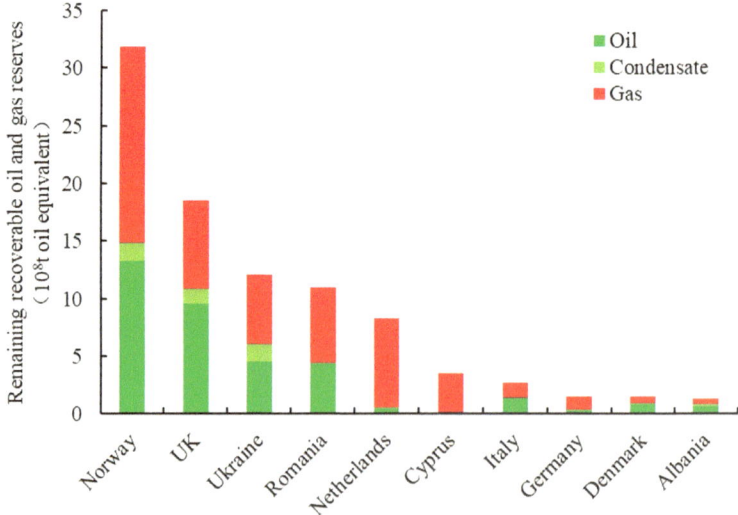

Fig. 6.1 Histogram of remaining recoverable reserves in major European countries (regions)

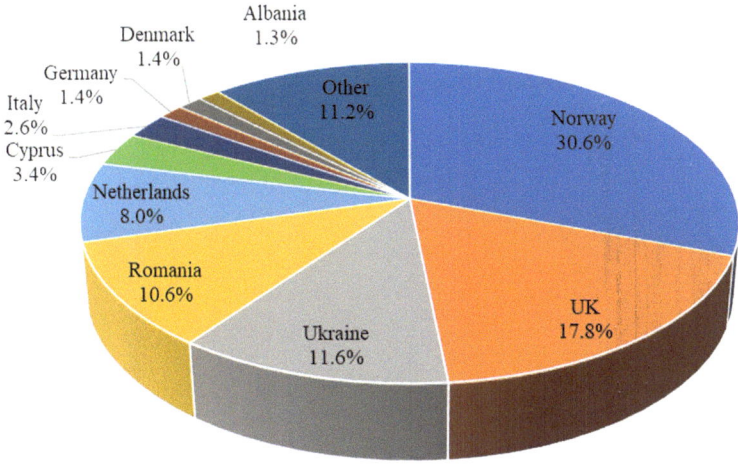

Fig. 6.2 Pie chart of remaining recoverable reserves in major European countries (regions)

North Carpathian Basin, Anglo-Dutch Basin, and Pannonian Basin (Fig. 6.3). The North Sea Basin, Dnieper-Donets Basin and Faroe-Shetland Basin are the top three basins with most remaining recoverable oil and gas reserves. The remaining recoverable reserves of oil and gas in these three basins account for 47.3% of the total remaining recoverable reserves in Europe (Fig. 6.4).

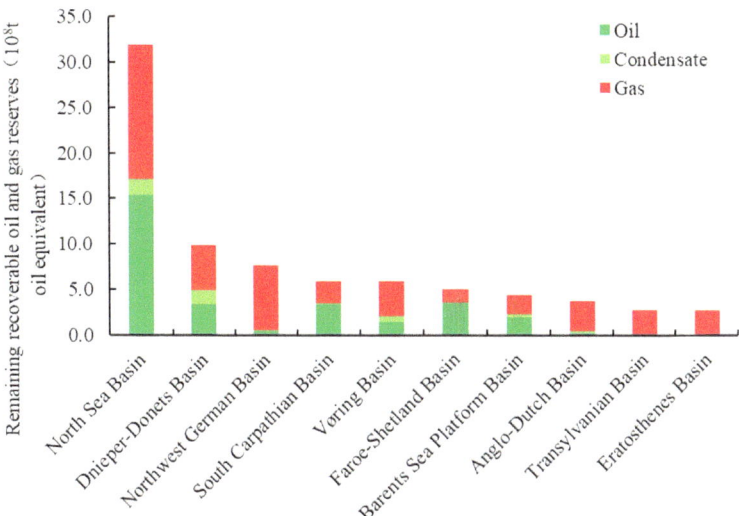

Fig. 6.3 Histogram of remaining recoverable reserves in major basins in Europe

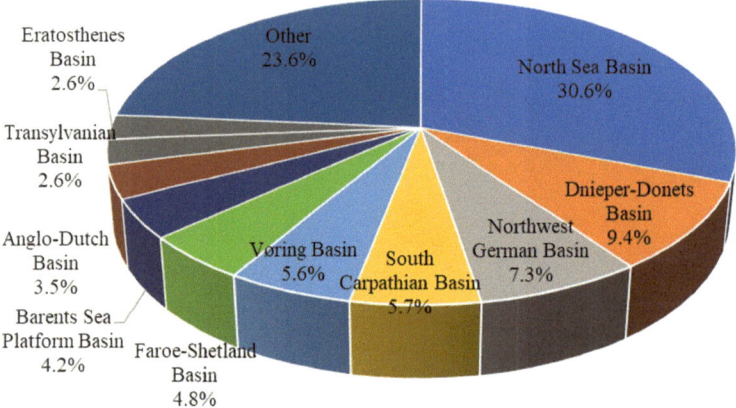

Fig. 6.4 Pie chart of remaining recoverable reserves in major basins in Europe

The remaining recoverable oil and gas reserves in the North Sea Basin amount to 31.9×10^8 t of oil equivalent, of which oil accounts for 48.2%, condensate accounts for 5.5%, and gas accounts for 46.3%. The remaining recoverable reserves in the Dnieper-Donets Basin are 9.8×10^8 t of oil equivalent, of which oil accounts for 35.1%, condensate accounts for 15.2%, and gas accounts for 49.7%. The remaining recoverable reserves in the Northwest German Basin are 7.6×10^8 t of oil equivalent, of which oil accounts for 6.1%, condensate accounts for 0.6%, and gas accounts for 93.3%.

6.1.1.3   Distribution of Remaining Recoverable Reserves in Onshore and Offshore Areas

The remaining recoverable oil and gas reserves in the onshore and offshore areas of Europe account for 34.3% and 65.7% of Europe's total remaining recoverable reserves, respectively (Fig. 6.5). The remaining recoverable reserves of onshore oil and gas are 35.7×10^8 t of oil equivalent, of which oil accounts for 39.0%, condensate accounts for 4.9%, and gas accounts for 56.1%. The remaining recoverable reserves of offshore oil and gas are 68.4×10^8 t of oil equivalent, of which oil accounts for 45.7%, condensate accounts for 5.7%, and gas accounts for 48.6%.

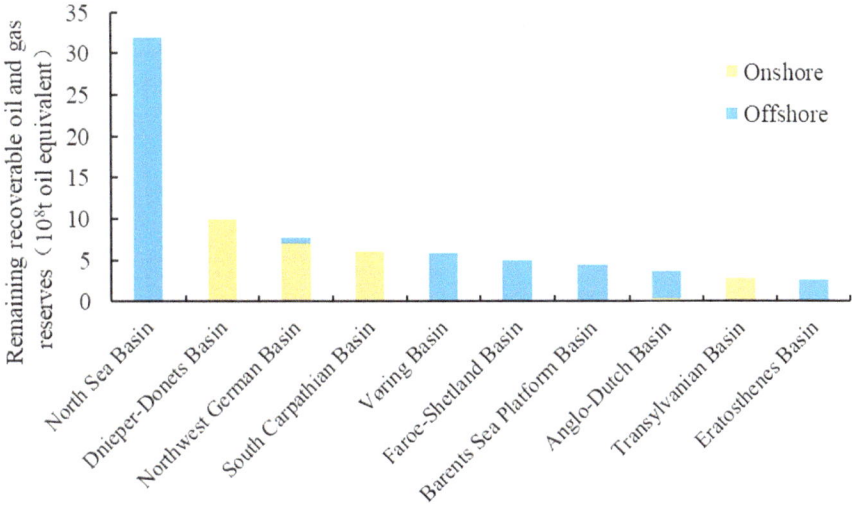

Fig. 6.5 Histogram of remaining recoverable reserves distributed in onshore and offshore in major basins in Europe

6.1.2 Trend of Reserve Growth from Discovered Oil and Gas Fields

The future reserve growth from discovered oil and gas fields in Europe is 32.3 × 10^8 t of oil equivalent, of which oil accounts for 27.1%, condensate accounts for 8.7%, and gas accounts for 64.2%.

6.1.2.1 Distribution of Reserve Growth in Discovered Oil and Gas Fields in Major Countries (Regions)

The future reserve growth from discovered oil and gas fields in Norway is 6.8 × 10^8 t of oil equivalent, accounting for 21.0% of the total reserve growth in Europe, of which oil accounts for 59.6%, condensate accounts for 18.0%, and gas accounts for 22.4%. The future reserve growth from discovered oil and gas fields in Poland are 6.1 × 10^8 t of oil equivalent, of which oil accounts for 2.4%, condensate accounts for 1.0%, and gas accounts for 96.6%. The future reserve growth from discovered oil and gas fields in UK are 5.9 × 10^8 t of oil equivalent, of which oil accounts for 35.7%, condensate accounts for 17.5%, and gas accounts for 46.8% (Figs. 6.6 and 6.7).

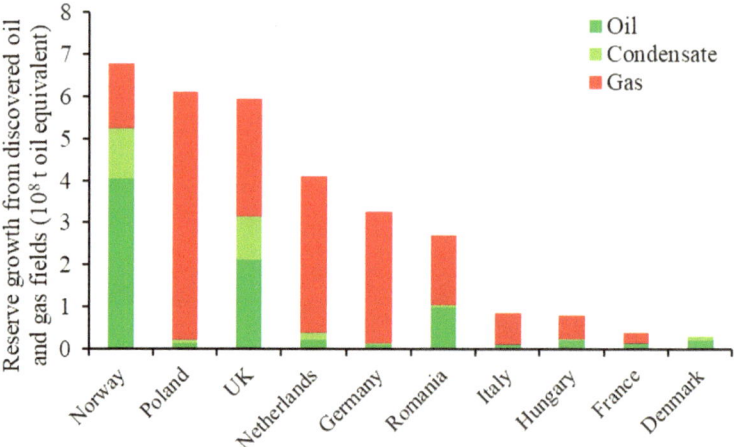

Fig. 6.6 Histogram of future reserve growth in discovered oil and gas fields in major European countries (regions)

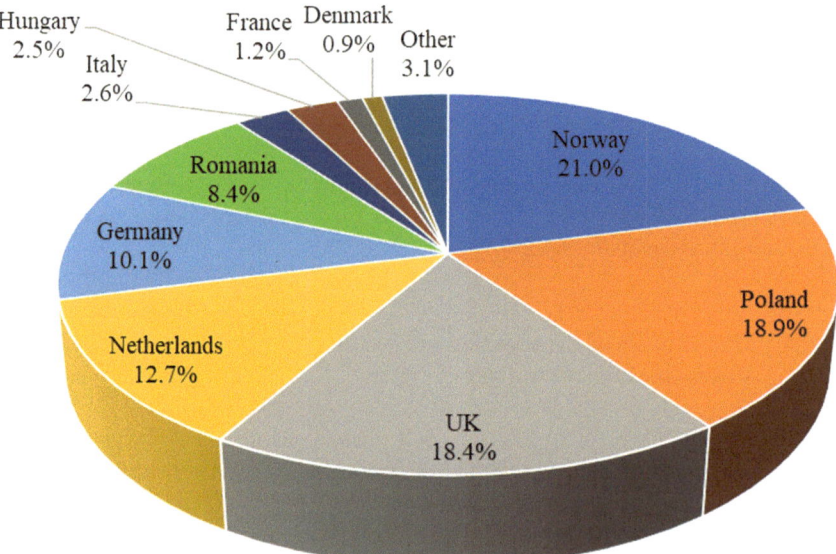

Fig. 6.7 Pie chart of future reserve growth in discovered oil and gas fields in major European countries (regions)

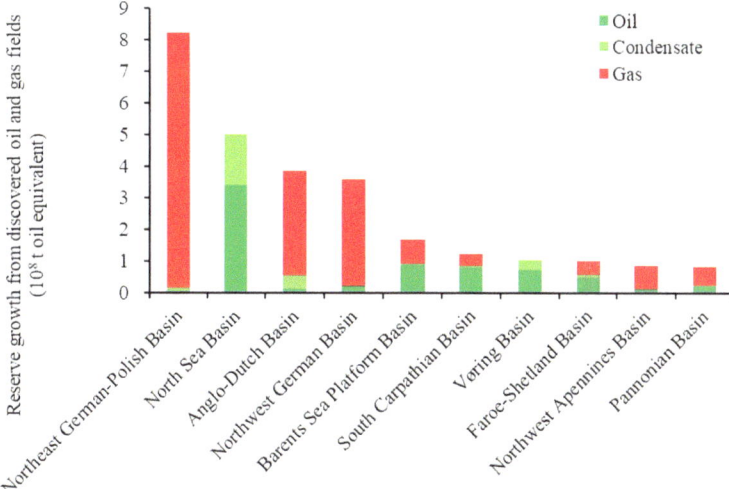

Fig. 6.8 Histogram of future reserve growth in discovered oil and gas fields in major basins in Europe

6.1.2.2 Distribution of Reserve Growth in Discovered Oil and Gas Fields in Major Basins

The Northeast German-Polish Basin, North Sea Basin, and Anglo-Dutch Basin rank among the top three in terms of future reserve growth from discovered oil and gas fields. The future reserve growth from discovered oil and gas fields in these three basins accounts for 52.8% of the total future reserve growth in Europe (Figs. 6.8 and 6.9). The future reserve growth from discovered oil and gas fields in the Northeast German-Polish Basin is 8.2×10^8 t of oil equivalent, of which oil accounts for 0.9%, condensate accounts for 1.0%, and gas accounts for 98.1%. The future reserve growth from discovered oil and gas fields in the North Sea Basin are 5.0×10^8 t of oil equivalent, of which oil accounts for 67.8% and condensate accounts for 32.2%. The future reserve growth from discovered oil and gas fields in the Anglo-Dutch Basin are 3.8×10^8 t of oil equivalent, of which oil accounts for 3.1%, condensate accounts for 11.2%, and gas accounts for 85.7%.

6.1.2.3 Distribution of Reserve Growth in Discovered Onshore and Offshore Oil and Gas Fields

In Europe, the distributions of future reserve growth in discovered onshore and offshore oil and gas fields are not uniform (Fig. 6.10). The future reserve growth from discovered onshore oil and gas fields in Europe is more than 16.4×10^8 t of oil equivalent, of which oil accounts for 27.8%, condensate accounts for 2.2%, and gas accounts for 70.0%. The future reserve growth from discovered offshore oil and gas

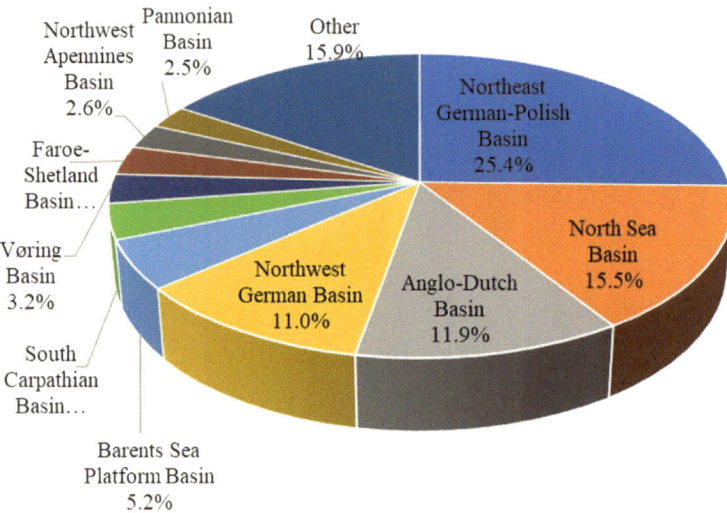

Fig. 6.9 Pie chart of future reserve growth in discovered oil and gas fields in major basins in Europe

fields in Europe are 15.9×10^8 t of oil equivalent, of which oil accounts for 51.3%, condensate accounts for 4.2%, and gas accounts for 44.5%.

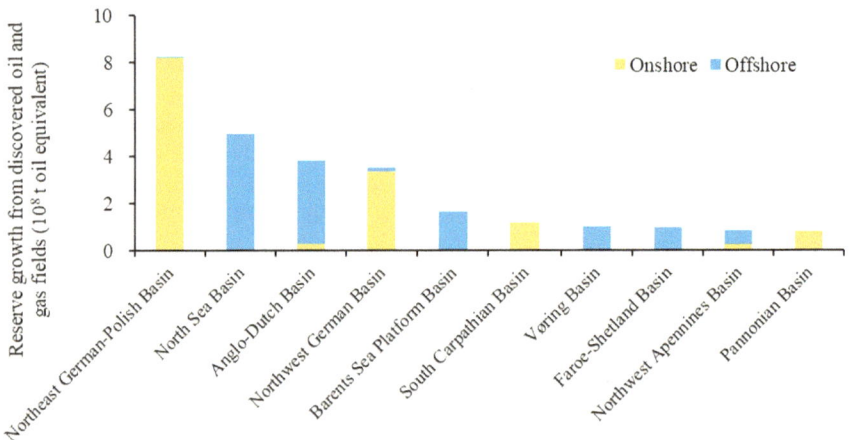

Fig. 6.10 Histogram of future reserve growth in discovered onshore and offshore oil and gas fields in major basins in Europe

6.1.3 Characteristics of Distributions of Undiscovered Oil and Gas Resources

The undiscovered oil and gas resources in Europe are 151.9×10^8 t of oil equivalent, of which oil accounts for 37.1%, condensate accounts for 9.0%, and gas accounts for 53.9%.

6.1.3.1 Distribution of Undiscovered Oil and Gas Resources in Major Countries (Regions)

The undiscovered oil and gas resources in Norway are 43.8×10^8 t of oil equivalent, of which oil accounts for 38.1%, condensate accounts for 14.3%, and gas accounts for 47.6%. The undiscovered oil and gas resources in Greenland are 27.9×10^8 t of oil equivalent, of which oil accounts for 45.8%, condensate accounts for 11.9%, and gas accounts for 42.3%. The undiscovered oil and gas resources in UK are 16.7 $\times 10^8$ t of oil equivalent, of which oil accounts for 57.7%, condensate accounts for 8.2%, and gas accounts for 34.1% (Figs. 6.11 and 6.12).

6.1.3.2 Distribution of Undiscovered Oil and Gas Resources in Major Basins

The East Greenland Basin, Vøring Basin, and Eratosthenes Basin rank among the top three in terms of potential undiscovered oil and gas resources. The discovered oil

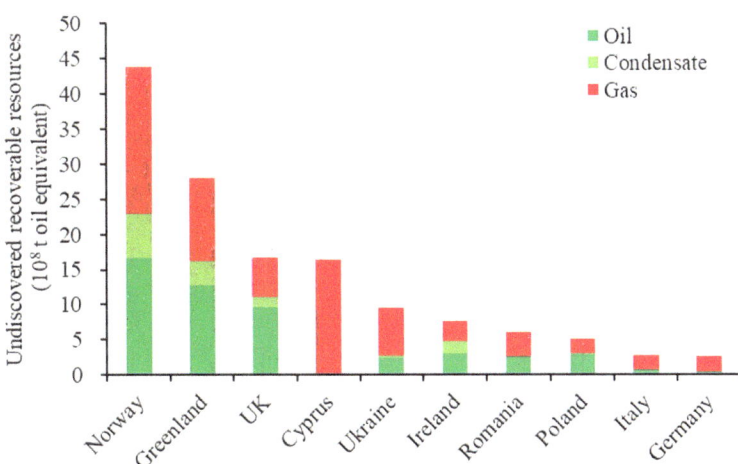

Fig. 6.11 Histogram of undiscovered recoverable oil and gas resources in major European countries (regions)

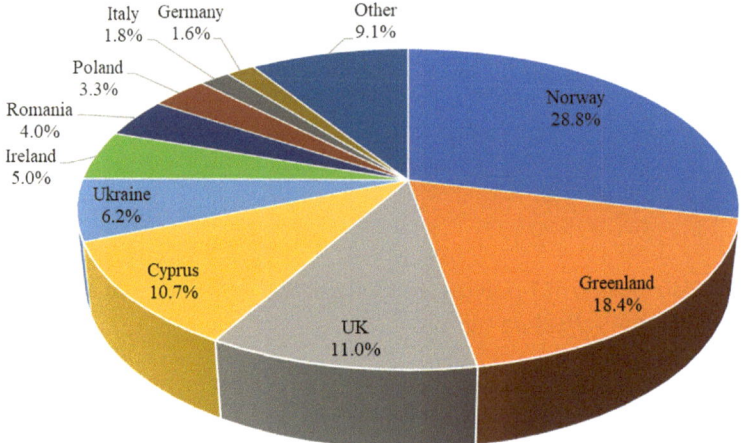

Fig. 6.12 Pie chart of undiscovered recoverable oil and gas resources in major European countries (regions)

and gas resources in these three basins account for 42.9% of the total undiscovered oil and gas resources in Europe (Figs. 6.13 and 6.14). The undiscovered oil and gas resources in the East Greenland Basin are 27.9×10^8 t of oil equivalent, of which oil accounts for 45.8%, condensate accounts for 11.9%, and gas accounts for 42.3%. The undiscovered oil and gas resources in the Vøring Basin are 20.8×10^8 t of oil equivalent, of which oil accounts for 35.5%, condensate accounts for 18.3%, and gas accounts for 46.2%. The undiscovered oil and gas resources in the Eratosthenes Basin are 16.3×10^8 t of oil equivalent, of which condensate accounts for 0.2% and gas accounts for 99.8%.

6.1.3.3 Distribution of Undiscovered Oil and Gas Resources in Onshore and Offshore Areas

The undiscovered offshore oil and gas resources in Europe are five times the undiscovered onshore resources. The undiscovered onshore oil and gas resources in Europe are 24.4×10^8 t of oil equivalent, of which oil accounts for 39.3%, condensate accounts for 4.8%, and gas accounts for 55.9% (Fig. 6.15). The undiscovered offshore oil and gas resources in Europe are 127.5×10^8 t of oil equivalent, of which oil accounts for 36.8%, condensate accounts for 9.8%, and gas accounts for 53.4%.

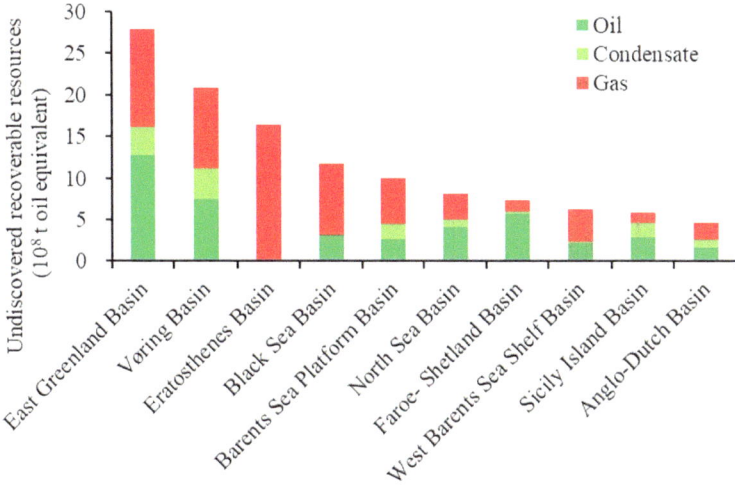

Fig. 6.13   Histogram of undiscovered recoverable oil and gas resources in major basins in Europe

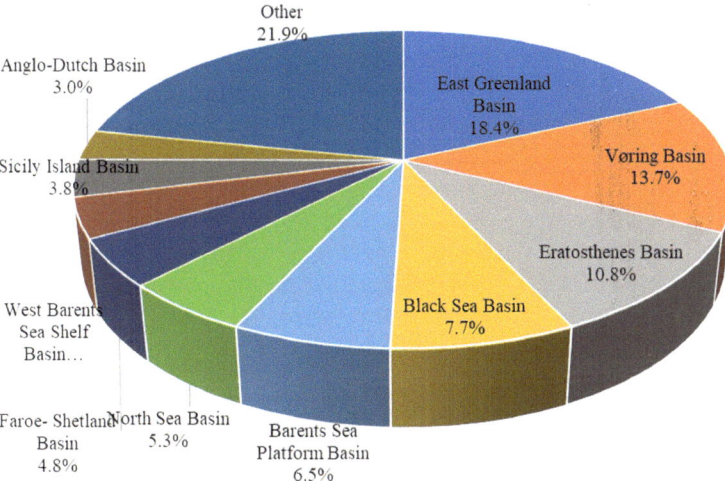

Fig. 6.14   Pie chart of undiscovered recoverable oil and gas resources in major basins in Europe

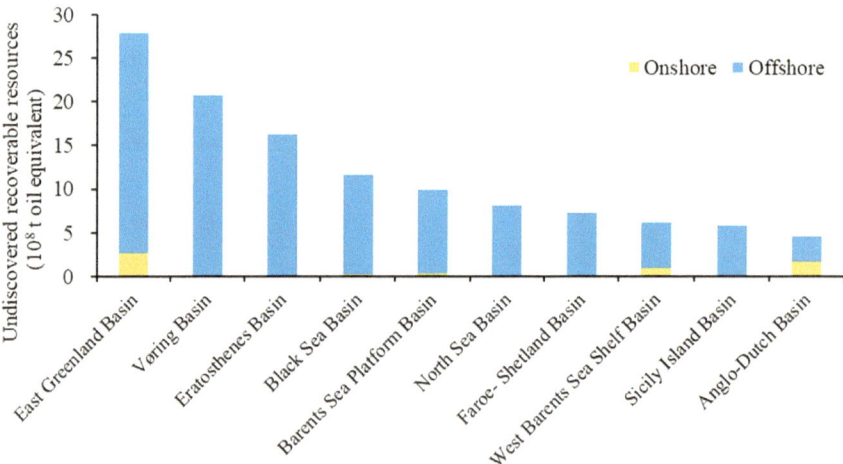

Fig. 6.15 Histogram of undiscovered recoverable oil and gas resources in onshore and offshore areas of major basins in Europe

6.2   Unconventional Oil and Gas Resources

The unconventional oil and gas resources in Europe include oil shale, heavy oil, shale oil, oil sands, shale gas, CBM, and tight gas. The total technically recoverable unconventional oil and gas resources in Europe are 517.6×10^8 t of oil equivalent, accounting for 7.1% of the global recoverable unconventional resources.

The technically recoverable unconventional oil resources in Europe are 325.9×10^8 t, accounting for 7.7% of the world's total technically recoverable unconventional oil resources. The technically recoverable resources of oil shale, heavy oil, shale oil,

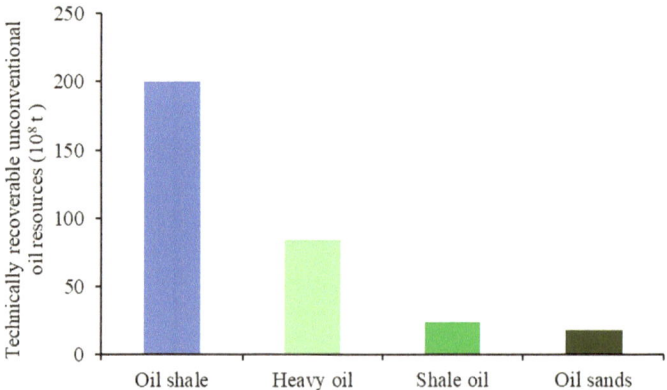

Fig. 6.16 Histogram of technically recoverable unconventional oil resources in Europe

Fig. 6.17 Pie chart of
technically recoverable
unconventional oil resources
in Europe

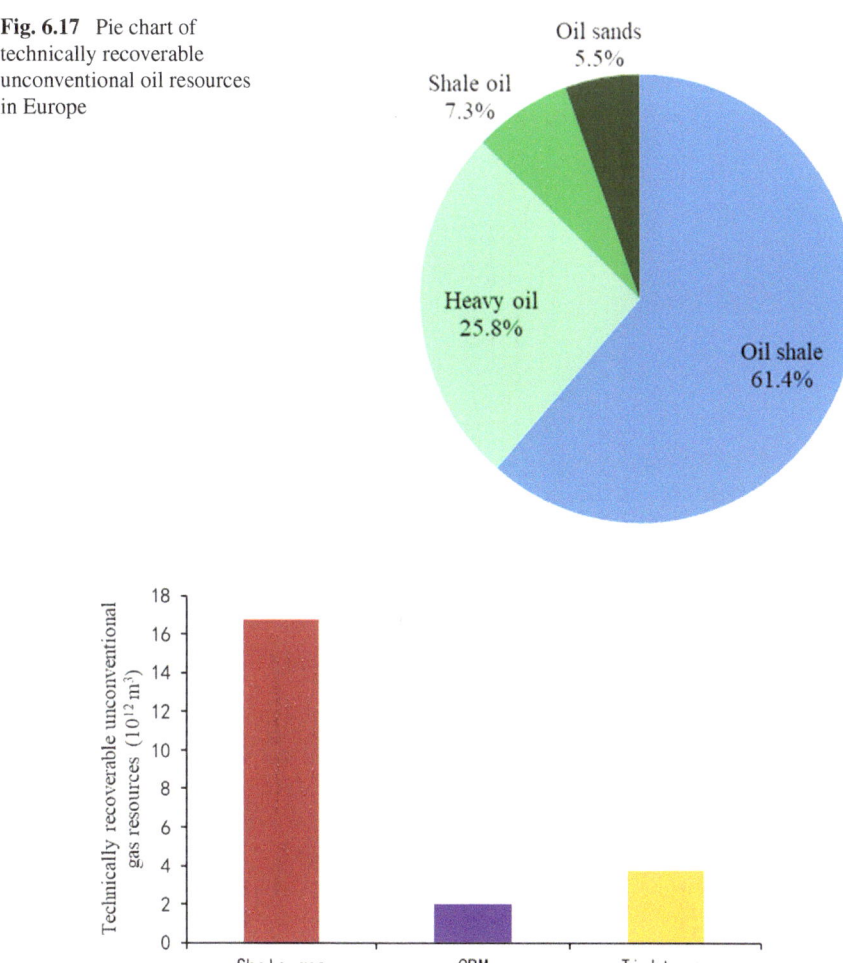

Fig. 6.18 Histogram of technically recoverable unconventional gas resources in Europe

and oil sands in Europe are 200.1×10^8 t, 84.2×10^8 t, 23.6×10^8 t, and 17.9×10^8 t, respectively (Figs. 6.16 and 6.17).

The technically recoverable unconventional gas resources in Europe are 22.4×10^{12} m^3, accounting for 6.4% of the world's total technically recoverable unconventional gas resources. The technically recoverable resources of shale gas, CBM, and tight gas in Europe are 16.7×10^{12} m^3, 2.0×10^{12} m^3, and 3.7×10^{12} m^3, respectively (Figs. 6.18 and 6.19).

Fig. 6.19 Pie chart of
technically recoverable
unconventional gas resources
in Europe

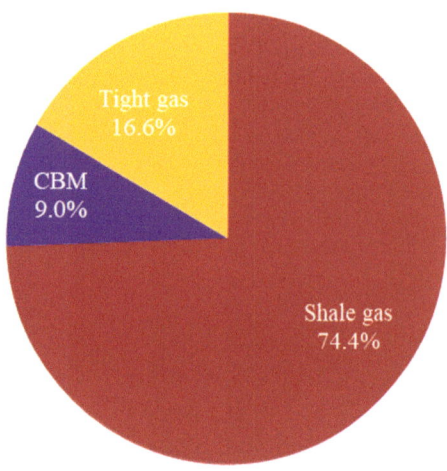

6.2.1 Distribution of Recoverable Unconventional Oil and Gas Resources in Major Countries (Regions)

6.2.1.1 Distribution of Recoverable Unconventional Oil Resources in Major Countries (Regions)

The unconventional oil resources in Europe are mainly distributed in three countries, namely, Ukraine, France, and Germany. The recoverable unconventional oil resources in Ukraine amount to 97.4×10^8 t, accounting for 29.9% of Europe's total recoverable unconventional oil resources and including 95.8×10^8 t of oil shale and 1.6×10^8 t of shale oil. The recoverable unconventional oil resources in France are 88.8×10^8 t, which mainly consist of oil shale and shale oil and include 78.7×10^8 t of oil shale and 6.6×10^8 t of shale oil. The recoverable unconventional oil resources in Germany are 38.7×10^8 t, which mainly consist of heavy oil, oil shale, and shale oil, including 35.5×10^8 t of heavy oil, 2.1×10^8 t of oil shale, and 1.0×10^8 t of shale oil (Figs. 6.20 and 6.21). In general, the unconventional oil resources in Ukraine and France are mainly oil shale, and the unconventional oil resources in Germany are mainly heavy oil resources.

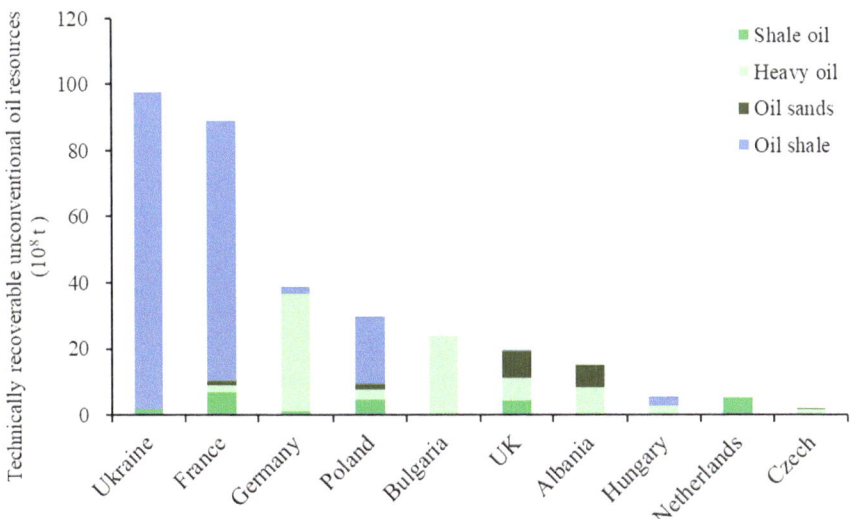

Fig. 6.20   Histogram of technically recoverable unconventional oil resources distributed in major European countries (regions)

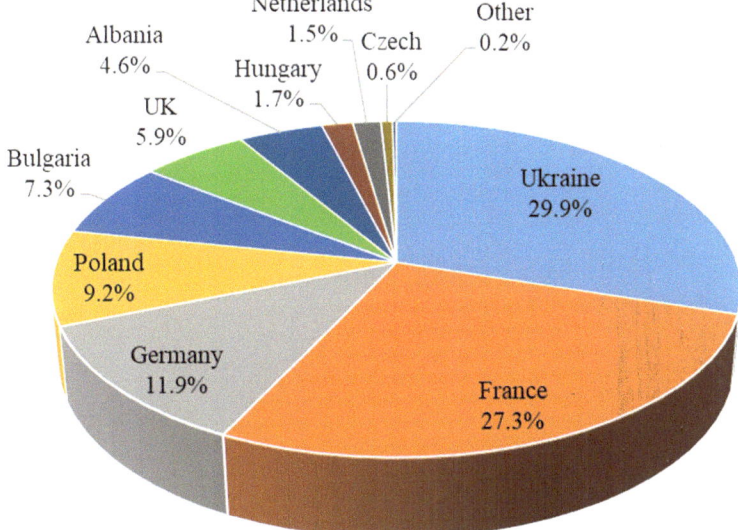

Fig. 6.21   Pie chart of technically recoverable unconventional oil resources distributed in major European countries (regions)

6.2.1.2   Distribution of Recoverable Unconventional Gas Resources in Major Countries (Regions)

The unconventional gas resources in Europe are mainly concentrated in Poland, Ukraine, and France. The recoverable unconventional gas resources in Poland are 5.1×10^{12} m^3, accounting for 22.6% of the total recoverable unconventional gas resources in Europe and including 4.2×10^{12} m^3 of shale gas and 0.4×10^{12} m^3 of CBM. The recoverable unconventional gas resources in Ukraine are 4.9×10^{12} m^3, including 2.2×10^{12} m^3 of shale gas, 1.6×10^{12} m^3 of CBM, and 1.1×10^{12} m^3 of tight gas. The recoverable unconventional gas resources in France are 3.7×10^{12} m^3, all of which are shale gas (Figs. 6.22 and 6.23).

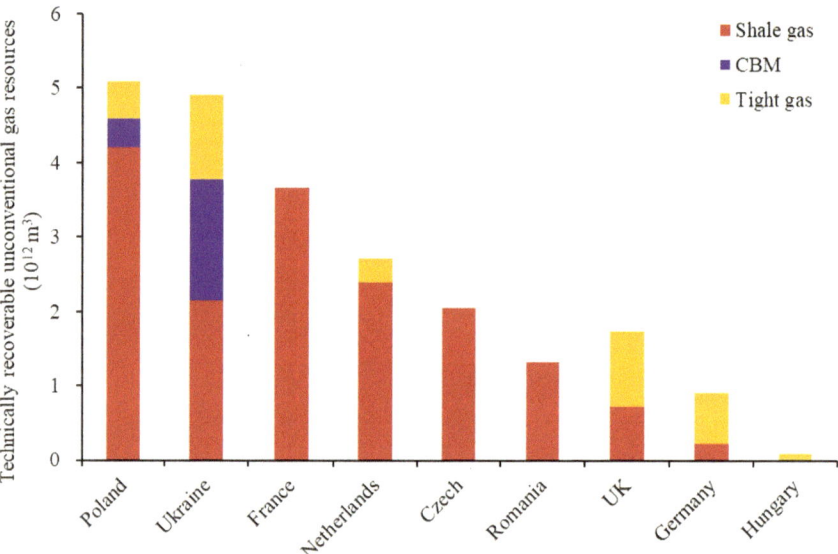

Fig. 6.22 Histogram of technically recoverable unconventional gas resources distributed in major European countries (regions)

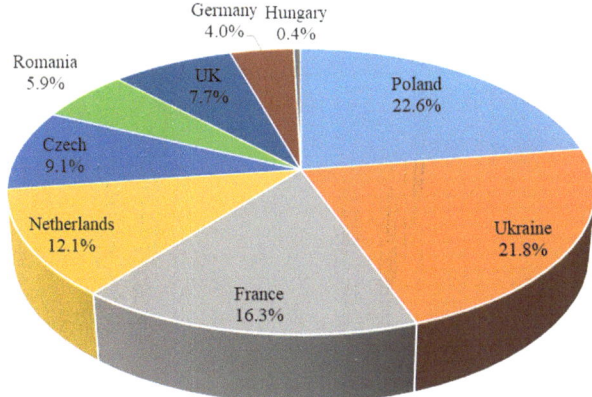

Fig. 6.23 Pie chart of technically recoverable unconventional gas resources distributed in major European countries (regions)

6.2.2 Distribution of Unconventional Oil and Gas Resources in Major Basins

6.2.2.1 Distribution of Unconventional Oil Resources in Major Basins

The unconventional oil resources in Europe are mainly concentrated in 15 basins, including the Dnieper-Donets Basin, Paris Basin, Baltic Sea Basin, and Pannonian Basin. In Europe, the Dnieper-Donets Basin ranks first in terms of total unconventional oil resources. The technically recoverable unconventional oil resources in the Dnieper-Donets Basin are 97.4×10^8 t, accounting for 29.9% of the total recoverable unconventional oil resources in Europe and including 95.8×10^8 t of oil shale and 1.6×10^8 t of shale oil. The technically recoverable unconventional oil resources in the Paris Basin rank second, which are 85.3×10^8 t, including 78.7×10^8 t of oil shale and 6.6×10^8 t of shale oil. The technically recoverable unconventional oil resources in the Northwest German Basin rank third, which are 36.5×10^8 t, including 35.5×10^8 t of heavy oil and 1.0×10^8 t of shale oil (Figs. 6.24 and 6.25).

The oil shale in Europe is mainly distributed in the Dnieper-Donets Basin, Paris Basin, Northwest German Basin, and Baltic Sea Basin. The heavy oil resources in Europe are mainly distributed in the Northwest German Basin, Mercia Platform Basin, Ionian Basin, North Sea Basin, North Carpathian Basin, Pannonian Basin, Molasse Basin, and South Carpathian Basin. The oil sands in Europe are mainly distributed in the North Sea Basin, Ionian Basin, North Carpathian Basin, Molasse Basin, and South Carpathian Basin. The shale oil resources in Europe are mainly distributed in the Anglo-Dutch Basin, Baltic Sea Basin, North Sea Basin, Dnieper-Donets Basin, Northwest German Basin, Wessex Basin, and Mercia Platform Basin.

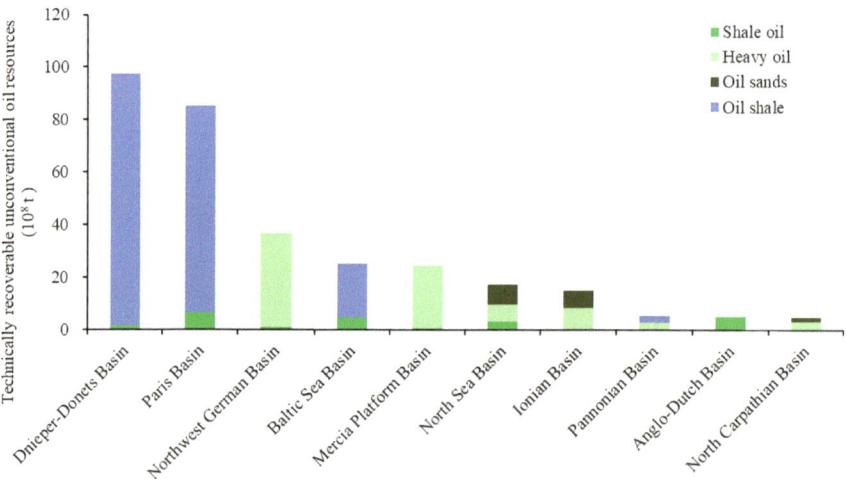

Fig. 6.24 Histogram of technically recoverable unconventional oil resources in top ten basins in Europe

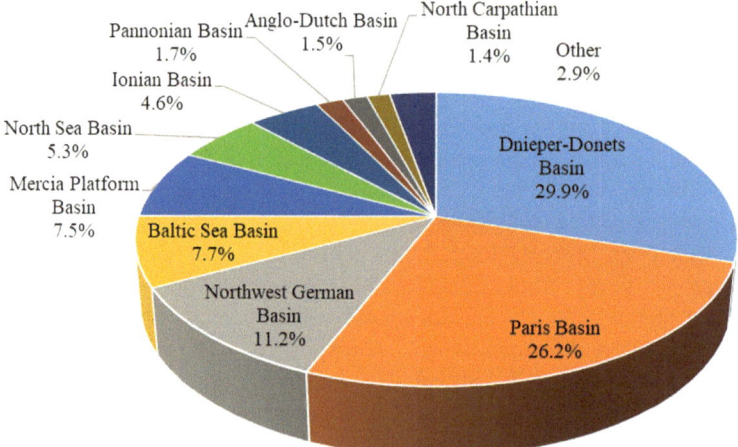

Fig. 6.25 Pie chart of technically recoverable unconventional oil resources distributed in major basins in Europe

6.2.2.2   Distribution of Unconventional Gas Resources in Major Basins

The unconventional gas resources in Europe are mainly concentrated in 10 basins, including the Dnieper-Donets Basin, Baltic Sea Basin, Paris Basin, Anglo-Dutch Basin, and South Carpathian Basin. The technically recoverable unconventional gas resources in the Dnieper-Donets Basin rank first in Europe, reaching 4.9×10^{12} m^3, which account for 21.8% of the total recoverable unconventional gas resources in

Europe and are mainly shale gas resources. The technically recoverable unconventional gas resources in the Baltic Sea Basin rank second, which are 4.2×10^{12} m^3, accounting for 18.7% of the total recoverable unconventional gas resources in Europe and mainly including shale gas. The technically recoverable unconventional gas resources in the Anglo-Dutch Basin rank third and amount to 3.7×10^{12} m^3, all of which are shale gas and tight gas (Figs. 6.26 and 6.27).

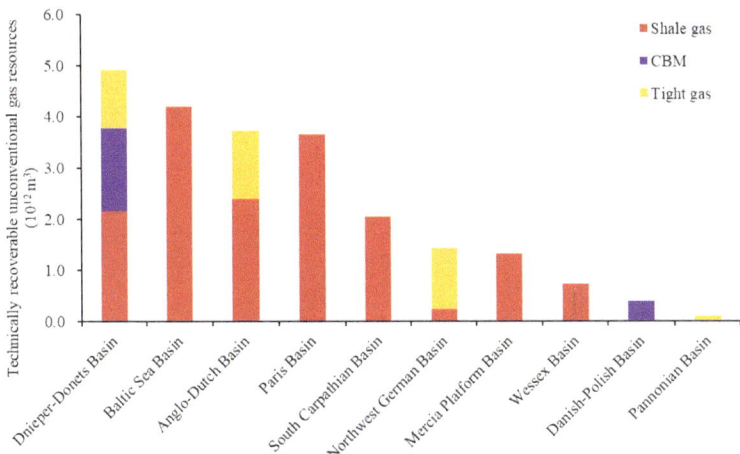

Fig. 6.26 Histogram of technically recoverable unconventional gas resources in top ten basins in Europe

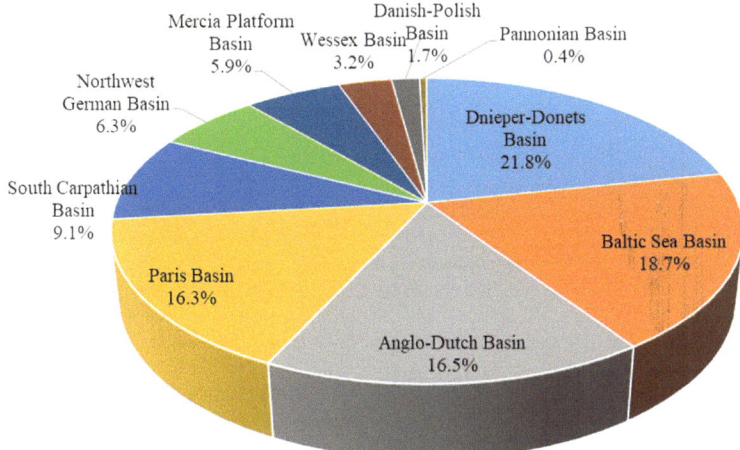

Fig. 6.27 Pie chart of technically recoverable unconventional gas resources in major basins in Europe

The shale gas resources in Europe are mainly distributed in the Baltic Sea Basin, Paris Basin, Anglo-Dutch Basin, Dnieper-Donets Basin, South Carpathian Basin, Mercia Platform Basin, Wessex Basin, and Northwest German Basin, among which the Baltic Sea Basin, Paris Basin and Anglo-Dutch Basin are the top three basins in terms of total shale gas resources. The recoverable shale gas resources in each of these basins are more than 2.4×10^{12} m^3. The CBM resources in Europe are mainly concentrated in the Dnieper-Donets Basin and Danish-Polish Basin, and the tight gas resources in Europe are distributed in the Anglo-Dutch Basin, Northwest German Basin, Dnieper-Donets Basin, and Pannonian Basin.

Chapter 7
Distribution of Oil and Gas Resources in Africa

Africa consists of 54 countries including Egypt, the Republic of South Africa, Kenya, and Nigeria, cover an area of 3020×10^4 km^2, and is the second largest continent worldwide. Seventy sedimentary basins are developed in Africa, cover an area of 3020×10^4 km^2. Onshore sedimentary basins have a total area of 2023.4×10^4 km^2 and mainly consist of cratonic and rift basins. Offshore sedimentary basins have a total area of 996.6×10^4 km^2 and mainly consist of passive margin basins. Africa has abundant oil and gas resources. 9.3% of the world's oil and gas resources are concentrated in Africa. The total recoverable oil and gas resources in Africa are 1740.0×10^8 tons of oil equivalent.

7.1 Conventional Oil and Gas Resources

The conventional oil and gas resources in Africa are 1151.0×10^8 tons of oil equivalent, accounting for 10.1% of the world's total conventional oil and gas resources. The recoverable conventional oil and gas resources in Africa are 585.0×10^8 tons, accounting for 9.0% of the world's total recoverable conventional resources. Africa's cumulative oil and gas production is 247.6×10^8 tons of oil equivalent, accounting for 9.5% of the global cumulative oil and gas production. The remaining recoverable oil and gas reserves in Africa are 337.4×10^8 tons of oil equivalent, accounting for 8.6% of the world's total remaining recoverable reserves. The predicted reserve growth from discovered oil and gas fields in the future is 165.6×10^8 tons of oil equivalent, accounting for 14.8% of the total global reserve growth from discovered oil and gas fields. The undiscovered recoverable oil and gas resources in Africa are 400.4×10^8 tons of oil equivalent, accounting for 10.5% of the world's total undiscovered recoverable resources.

© The Author(s) 2024
L. Dou et al., *Global Oil and Gas Resources: Potential and Distribution*,
https://doi.org/10.1007/978-981-97-4756-6_7

7.1.1   Distribution of Remaining Recoverable Reserves

The remaining recoverable reserves of oil and gas in Africa are 337.4×10^8 tons of oil equivalent, including 122.2×10^8 tons of oil, accounting for 36.2%, 20.0×10^8 tons of condensate, accounting for 5.9%, and 22.9×10^{12} m^3 of gas, accounting for 57.9%.

7.1.1.1   Distribution of Remaining Recoverable Reserves in Major Countries (Regions)

The remaining recoverable reserves of oil and gas in Africa are mainly distributed in Nigeria, Algeria, Libya, Mozambique, Angola, and Egypt. The remaining recoverable reserves in these countries account for 77.7% of the total remaining recoverable reserves in Africa (Figs. 7.1 and 7.2). The remaining recoverable reserves of oil and gas in Nigeria are 86.4×10^8 tons of oil equivalent, of which oil accounts for 42.8%, condensate accounts for 5.7%, and gas accounts for 51.5%. The remaining recoverable reserves of oil and gas in Algeria and Libya are 50.8×10^8 tons of oil equivalent and 37.8×10^8 tons of oil equivalent, respectively, ranking second and third in Africa. Among the remaining recoverable reserves in these two countries, oil reserves account for 30.5% and 54.9%, condensate reserves account for 9.6% and 8.8%, and gas reserves account for 59.9% and 36.3%, respectively.

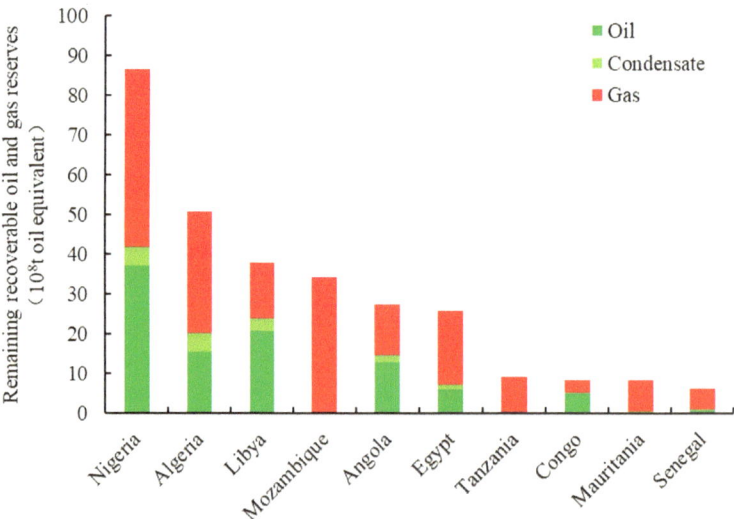

Fig. 7.1 Histogram of remaining recoverable reserves in major African countries (regions)

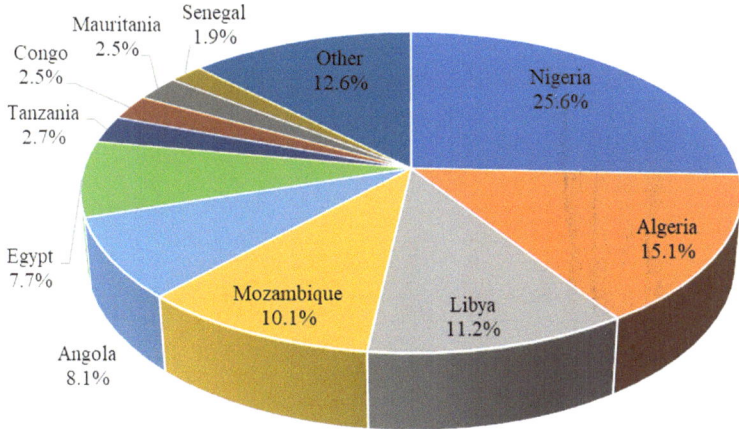

Fig. 7.2 Pie chart of remaining recoverable reserves in major African countries (regions)

7.1.1.2 Distribution of Remaining Recoverable Reserves in Major Basins

The remaining recoverable reserves of oil and gas in Africa are mainly distributed in the Niger Delta Basin, Ghadames Basin, Rovuma Basin, Lower Congo Basin, Sirte Basin, and Senegal Basin. The remaining recoverable reserves of oil and gas in these basins account for 69.5% of the total remaining recoverable reserves of all basins in Africa (Figs. 7.3 and 7.4). The remaining recoverable oil and gas reserves in the Niger Delta Basin are 90.0×10^8 tons of oil equivalent, of which oil accounts for 42.1%, condensate accounts for 5.9%, and gas accounts for 52.0%. The remaining recoverable reserves of oil and gas in the Ghadames Basin and Rovuma Basin are 41.3×10^8 tons of oil equivalent and 35.3×10^8 tons of oil equivalent, ranking second and third in Africa, of which oil accounts for 39.1% and 0%, condensate accounts for 8.3% and 0.9%, and gas accounts for 52.6% and 99.1%, respectively.

7.1.1.3 Distribution of Remaining Recoverable Reserves in Onshore and Offshore Areas

The distributions of remaining recoverable oil and gas reserves in onshore and offshore areas of Africa are uneven. The remaining recoverable oil and gas reserves in onshore and offshore areas account for 44.3% and 55.7%, respectively (Fig. 7.5). In offshore areas of Africa, the remaining recoverable gas reserves are more than the remaining recoverable oil reserves. In onshore areas, the remaining recoverable oil reserves are more than the remaining recoverable gas reserves.

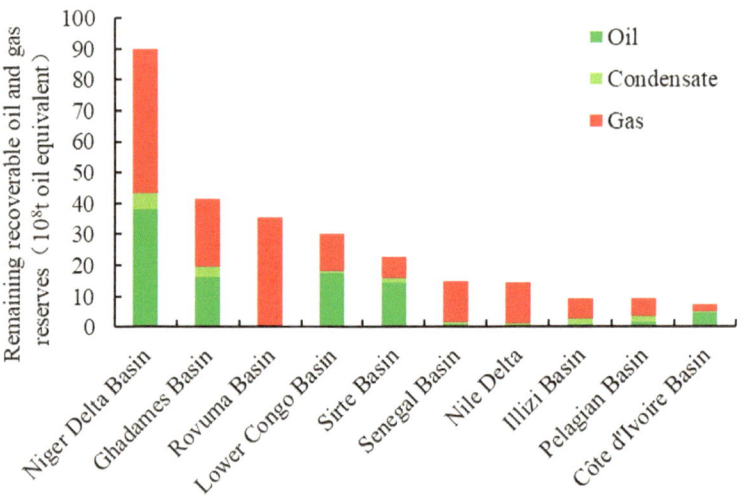

Fig. 7.3 Histogram of remaining recoverable reserves distributed in major basins in Africa

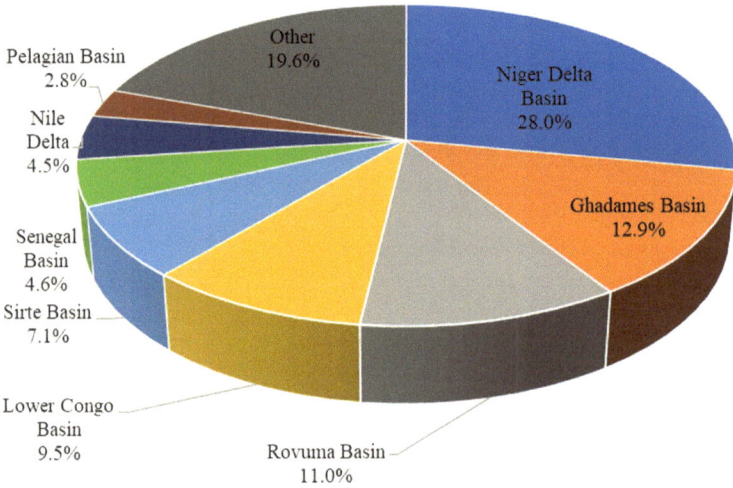

Fig. 7.4 Pie chart of remaining recoverable reserves distributed in major basins in Africa

The remaining recoverable reserves of oil and gas in offshore areas are 187.9×10^8 tons of oil equivalent, of which oil, condensate and gas account for 30.8%, 5.4% and 63.8%, respectively. The remaining recoverable reserves in onshore areas are 149.6×10^8 tons of oil equivalent, of which oil, condensate and gas account for 45.3%, 6.2% and 48.5%, respectively.

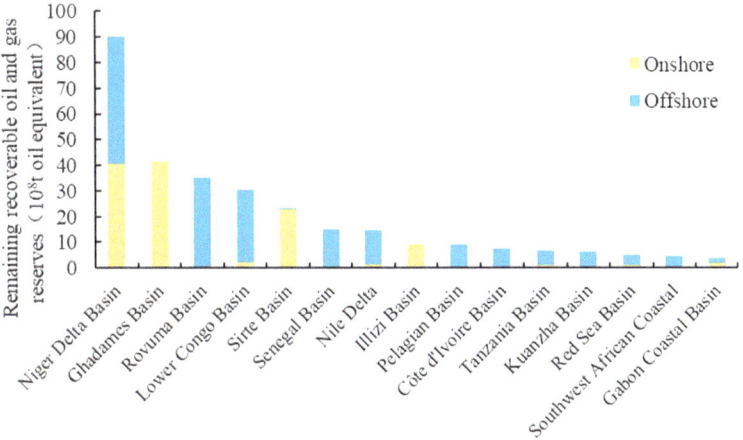

Fig. 7.5 Histogram of remaining recoverable reserves distributed in onshore and offshore areas of major basins in Africa

7.1.2 Trend of Reserve Growth from Discovered Oil and Gas Fields

The reserve growth in discovered oil and gas fields in Africa is 165.6×10^8 tons of oil equivalent, including 76.1×10^8 tons of oil, accounting for 45.9%, 5.2×10^8 tons of condensate, accounting for 3.1%, and 9.9×10^{12} m^3 of gas, accounting for 51.0%.

7.1.2.1 Distribution of Reserve Growth in Discovered Oil and Gas Fields in Major Countries (Regions)

The reserve growth in discovered oil and gas fields in Africa is mainly distributed in Egypt, Mozambique, Tanzania, Nigeria, the Republic of Congo, Libya, and Algeria. The reserve growth in these countries account for 92.6% of the total reserve growth in Africa. Egypt has the largest amounts of reserve growth (Figs. 7.6 and 7.7).

The future reserve growth of discovered oil and gas fields in Egypt is 35.4×10^8 tons of oil equivalent, of which oil accounts for 67.3%, condensate accounts for 5.5%, and gas accounts for 27.2%. The future reserve growth of discovered oil and gas fields in Mozambique and Tanzania rank second and third in Africa and are 33.2×10^8 tons of oil equivalent and 19.2×10^8 tons of oil equivalent, respectively, of which condensate accounts for 1.5% and 0.1% and gas accounts for 98.5% and 99.9%, respectively.

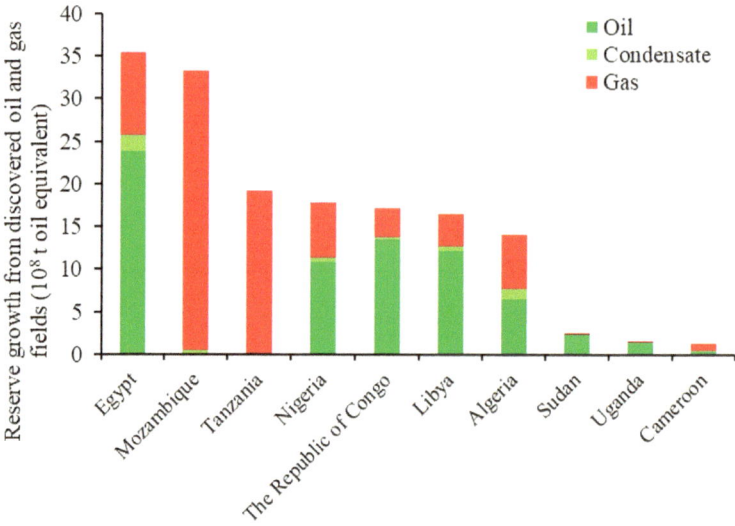

Fig. 7.6 Histogram of reserve growth in discovered oil and gas fields of major African countries (regions)

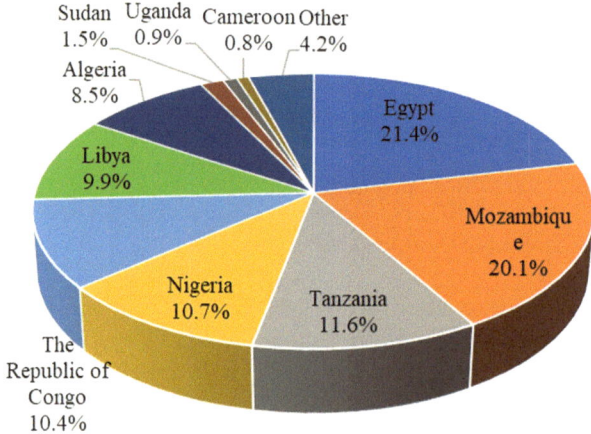

Fig. 7.7 Pie chart of reserve growth in discovered oil and gas fields of major African countries (regions)

7.1.2.2   Distribution of Reserve Growth in Discovered Oil and Gas Fields in Major Basins

The reserve growth in discovered oil and gas fields in Africa is mainly distributed in the Rovuma Basin, Nile Delta Basin, Tanzanian Basin, Niger Delta Basin, Lower Congo Basin, Sirte Basin, and Illizi Basin. The reserve growth in these basins accounts for 83.9% of the total reserve growth in Africa. The Rovuma Basin has the largest amounts of reserve growth (Figs. 7.8 and 7.9).

The future reserve growth in the Rovuma Basin is 33.2×10^8 tons of oil equivalent, of which condensate accounts for 1.4% and gas accounts for 98.6%. The future reserve growth in the Nile Delta Basin and Tanzanian Basin rank second and third in Africa and are 28.7×10^8 tons of oil equivalent and 19.2×10^8 tons of oil equivalent,

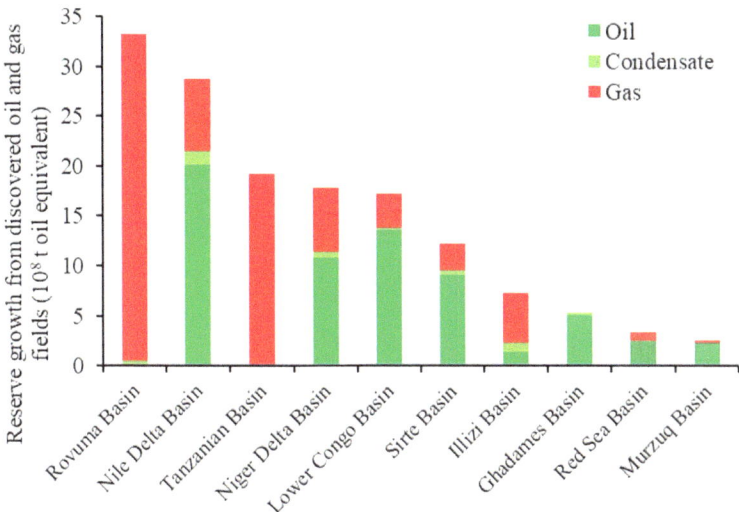

Fig. 7.8 Histogram of reserve growth in discovered oil and gas fields of major basins in Africa

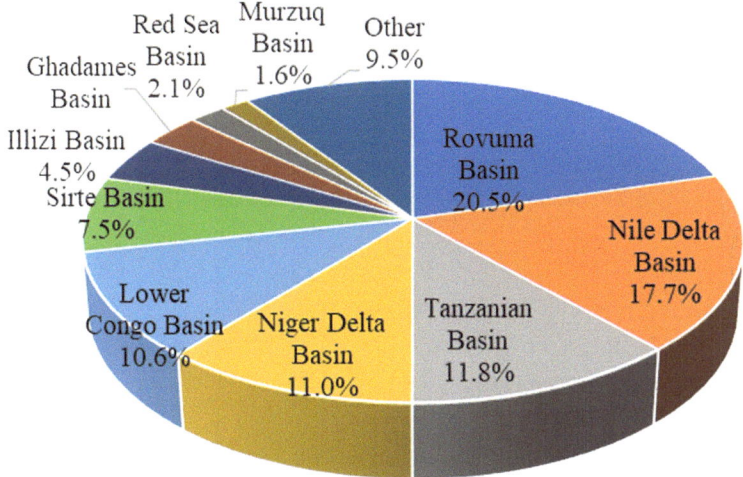

Fig. 7.9 Pie chart of reserve growth in discovered oil and gas fields of major basins in Africa

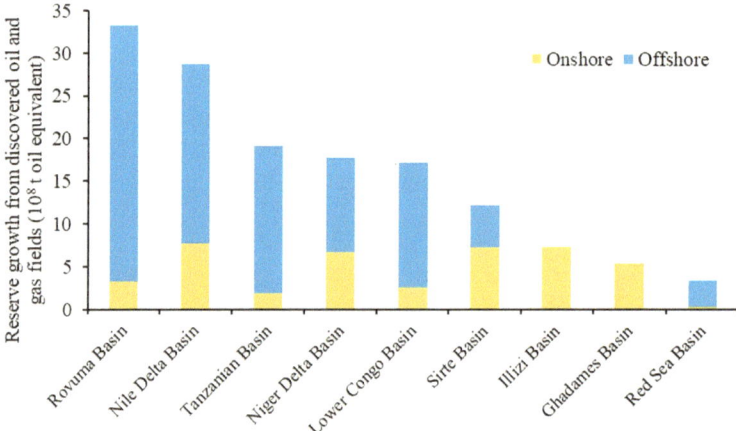

Fig. 7.10 Histogram of future reserve growth in discovered oil and gas fields of major basins in Africa

respectively, of which condensate accounts for 70.2% and 0.3%, condensate account for 4.8% and 0.1%, and gas accounts for 25.0% and 99.6%, respectively.

7.1.2.3 Distribution of Reserve Growth in Discovered Onshore and Offshore Oil and Gas Fields

The additional oil and gas reserves in Africa are unevenly distributed in offshore and onshore areas. The additional oil and gas reserves in offshore and onshore areas account for 65.4% and 34.6%, respectively (Fig. 7.10). In offshore areas of Africa, the additional gas reserves are more than the additional oil reserves. In onshore areas, the additional oil reserves are more than the additional gas reserves.

The future reserve growth in offshore areas of Africa is 108.4×10^8 tons of oil equivalent, of which oil, condensate and gas accounts for 39.2%, 2.4% and 58.4%, respectively. The future reserve growth in onshore areas of Africa are 57.2×10^8 tons of oil equivalent, of which oil, condensate and gas accounts for 58.6%, 4.6% and 36.8%, respectively.

7.1.3 Characteristics of Distributions of Undiscovered Oil and Gas Resources

The total undiscovered oil and gas resources in Africa are 400.4×10^8 tons of oil equivalent, including 157.1×10^8 tons of oil, accounting for 39.2%, 47.4×10^8 tons of condensate, accounting for 11.8%, and 22.9×10^{12} m^3 of gas, accounting for 48.9%.

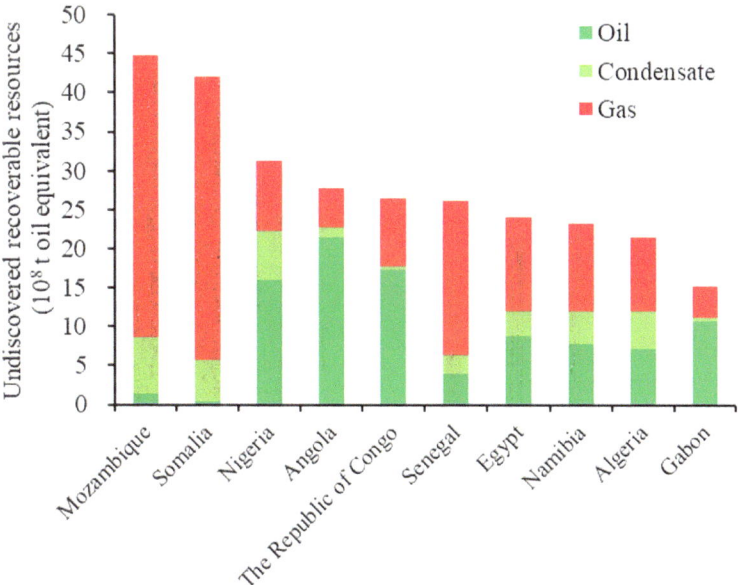

Fig. 7.11 Histogram of undiscovered recoverable oil and gas resources in major African countries (regions)

7.1.3.1 Distribution of Undiscovered Oil and Gas Resources in Major Countries (Regions)

The undiscovered oil and gas resources in Africa are mainly distributed in Mozambique, Somalia, Nigeria, Angola, the Republic of Congo, and Senegal. The undiscovered oil and gas resources in these countries account for 49.5% of the total undiscovered oil and gas resources in Africa. Mozambique has the largest amounts of undiscovered oil and gas resources (Figs. 7.11 and 7.12). The undiscovered oil and gas resources in Mozambique are 44.7×10^8 tons of oil equivalent, of which gas is dominant and accounts for 80.7%. The undiscovered oil and gas resources in Somalia and Nigeria rank second and third in Africa and are 42.0×10^8 tons of oil equivalent and 31.2×10^8 tons of oil equivalent, respectively, of which oil accounts for 1.1% and 51.3%, condensate accounts for 12.3% and 20.0%, and gas accounts for 86.6% and 28.7%, respectively.

7.1.3.2 Distribution of Undiscovered Oil and Gas Resources in Major Basins

The undiscovered oil and gas resources in Africa are mainly distributed in the Somali Deep Sea Basin, Niger Delta Basin, Mozambique Basin, Southwest African Costal Basin, Lower Congo Basin, Kwanza Basin, Senegal Basin, Rovuma Basin, Gabon

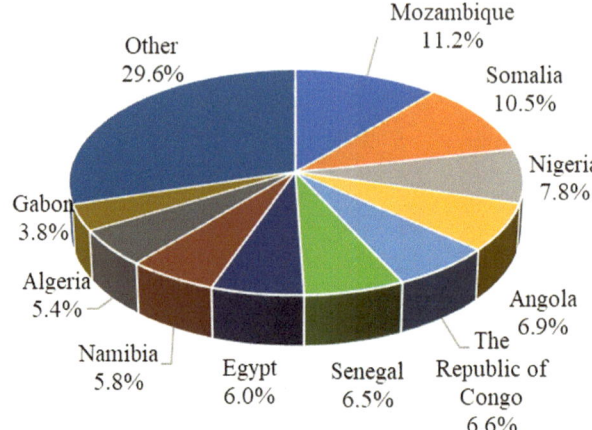

Fig. 7.12 Pie chart of undiscovered recoverable oil and gas resources in major African countries (regions)

Coastal Basin, Tanzanian Basin, and so on. The undiscovered oil and gas resources in these basins account for 66.3% of the total undiscovered oil and gas resources in Africa. The Somali Deep Sea Basin has the largest amounts of undiscovered oil and gas resources (Figs. 7.13 and 7.14).

The undiscovered oil and gas resources in the Somali Deep Sea Basin are 35.5 $\times 10^8$ tons of oil equivalent, of which oil accounts for 1.3%, condensate accounts

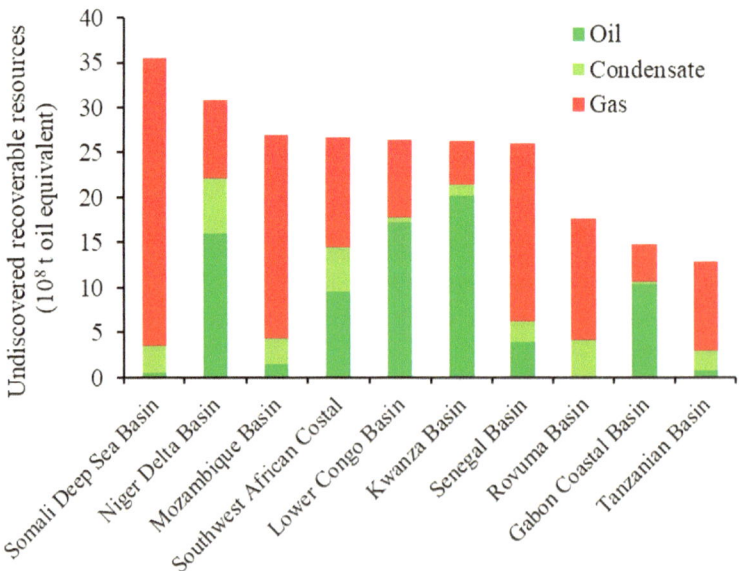

Fig. 7.13 Histogram of undiscovered oil and gas resources in major basins in Africa

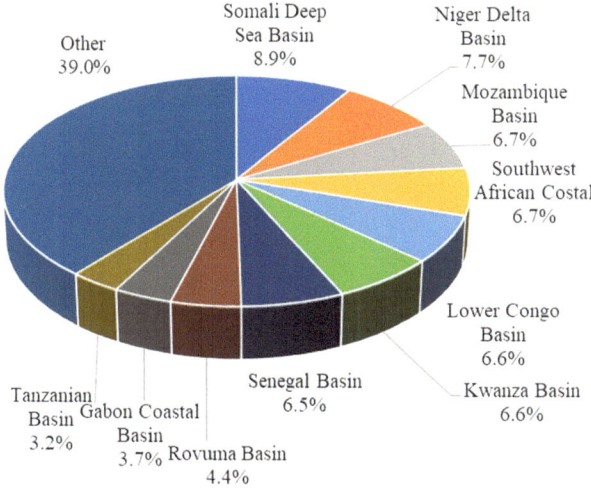

Fig. 7.14 Pie chart of undiscovered oil and gas resources in major basins in Africa

for 8.5%, and gas accounts for 90.2%. The undiscovered oil and gas resources in the Niger Delta Basin and Mozambique Basin rank second and third in Africa and are 30.8×10^8 tons of oil equivalent and 27.0×10^8 tons of oil equivalent, respectively, of which oil accounts for 52.0% and 5.3%, condensate accounts for 20.3% and 11.0%, and gas accounts for 27.7% and 83.7%, respectively.

7.1.3.3 Distribution of Undiscovered Oil and Gas Resources in Onshore and Offshore Areas

The undiscovered oil and gas resources in Africa are unevenly distributed in offshore and onshore areas. The undiscovered oil and gas resources in offshore and onshore areas account for 68.9% and 31.1%, respectively (Fig. 7.15). In onshore areas of Africa, the undiscovered oil resources are more than the undiscovered gas resources. In offshore areas, the undiscovered gas resources are more than the undiscovered oil resources.

The undiscovered onshore oil and gas resources in Africa are 124.7×10^8 tons of oil equivalent, of which oil accounts for 49.7%, condensate accounts for 10.7%, and gas accounts for 39.6%. The undiscovered offshore oil and gas resources in Africa are 275.7×10^8 tons of oil equivalent, of which oil accounts for 30.4%, condensate accounts for 10.8%, and gas accounts for 58.8%.

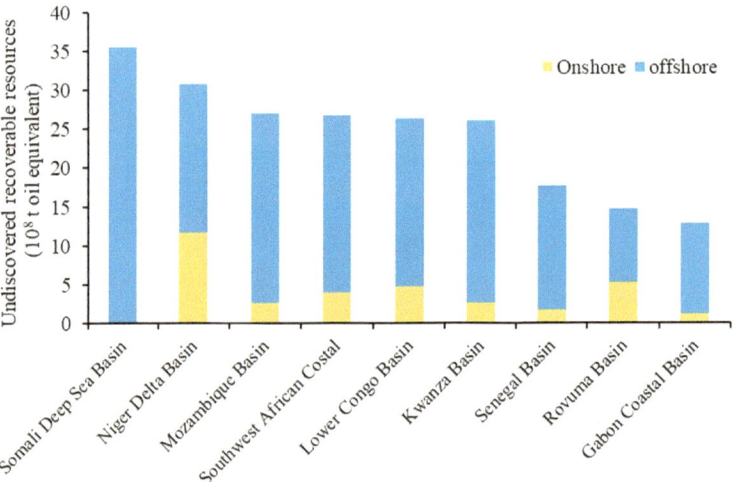

Fig. 7.15 Histogram of undiscovered recoverable oil and gas resources in onshore and offshore areas of major basins in Africa

7.2   Unconventional Oil and Gas Resources

The technically recoverable unconventional oil and gas resources in Africa are 589.0 $\times 10^8$ tons of oil equivalent, which account for 8.1% of the world's total recoverable unconventional resources and mainly consist of oil shale, heavy oil, shale oil, oil sands, and shale gas.

The recoverable unconventional oil resources in Africa are 222.8 $\times 10^8$ tons of oil equivalent, which account for 5.2% of the world's total recoverable unconventional oil resources and include 69.7 $\times 10^8$ tons of oil shale, accounting for 31.3%, 64.8 $\times 10^8$ tons of heavy oil, accounting for 29.1%, 63.3 $\times 10^8$ tons of shale oil, accounting for 28.4%, and 25.0 $\times 10^8$ tons of oil sands, accounting for 11.2% (Figs. 7.16 and 7.17).

The recoverable unconventional gas resources in Africa are 42.9 $\times 10^{12}$ m^3, which account for 12.2% of the world's total recoverable unconventional gas resources and mainly consist of shale gas (Figs. 7.18 and 7.19).

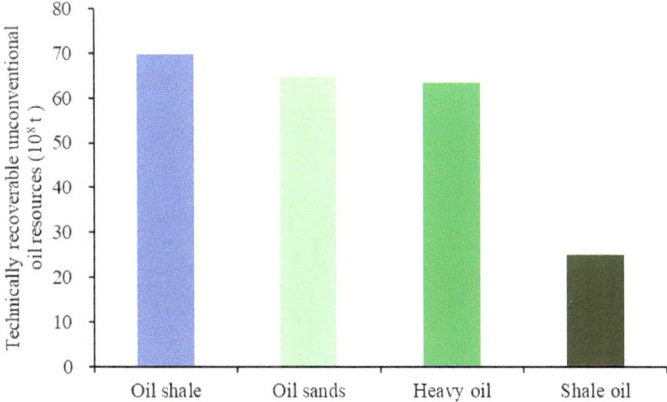

Fig. 7.16 Histogram of technically recoverable unconventional oil resources in Africa

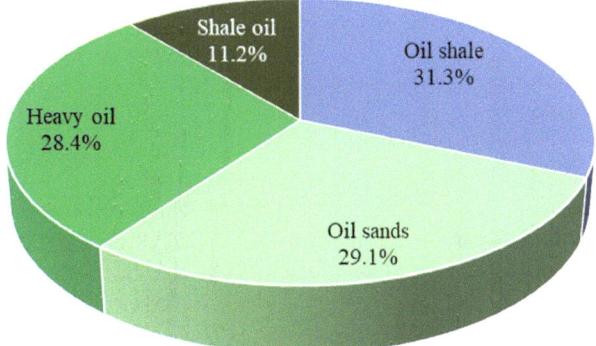

Fig. 7.17 Pie chart of technically recoverable unconventional oil resources in Africa

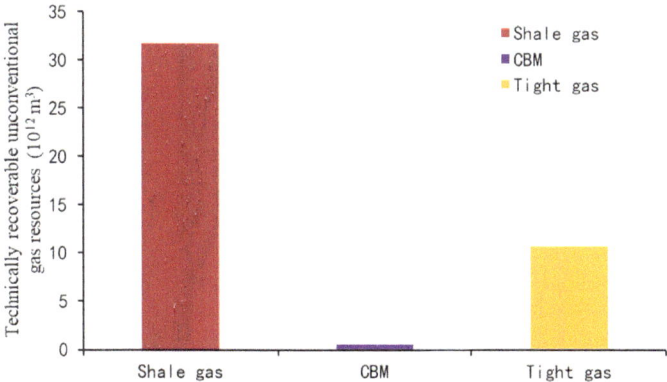

Fig. 7.18 Histogram of technically recoverable unconventional gas resources in Africa

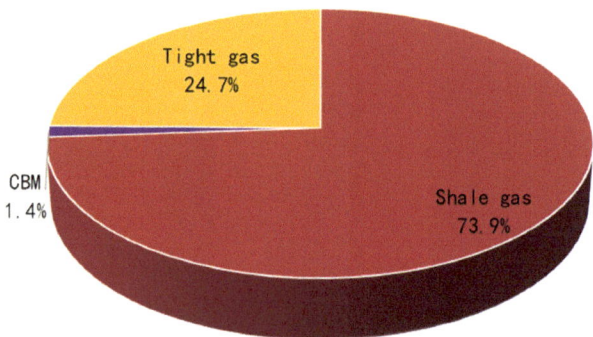

Fig. 7.19 Pie chart of technically recoverable unconventional gas resources in Africa

7.2.1   *Distribution of Recoverable Unconventional Oil and Gas Resources in Major Countries (Regions)*

7.2.1.1   Distribution of Undiscovered Oil Resources in Major Countries (Regions)

The unconventional oil and gas resources in Africa are mainly distributed in 14 countries, including Algeria, the Republic of South Africa, Morocco, Egypt, Libya, Madagascar, Nigeria, Chad, the Republic of Congo, Niger, Sudan, Angola, Senegal, and Benin, and mainly consist of oil shale in Morocco, heavy oil in Madagascar, and shale oil in Libya. The recoverable oil shale in Morocco is 67.4×10^8 tons, accounting for 30.3% of the total recoverable unconventional oil resources in Africa; the recoverable heavy oil resources in Madagascar are 34.8×10^8 tons, accounting for 15.6%; the recoverable shale oil resources in Libya are 27.4×10^8 tons, accounting for 12.3%; the recoverable unconventional oil resources in Algeria amount to 19.0×10^8 tons and mainly consist of shale oil; and the recoverable unconventional oil resources in Egypt amount to 22.7×10^8 tons and mainly consist of heavy oil.

In Africa, oil shale are mainly distributed in Morocco and Nigeria; heavy oil resources are mainly distributed in Madagascar, Egypt, and the Republic of Congo; shale oil resources are mainly distributed in Libya, Algeria, Egypt, and Chad; and oil sands are mainly concentrated in Nigeria (Figs. 7.20 and 7.21).

7.2.1.2   Distribution of Undiscovered Gas Resources in Major Countries (Regions)

The unconventional gas resources in Africa are mainly concentrated in four countries, namely, Algeria, the Republic of South Africa, Egypt, and Libya. Shale gas is the main type of unconventional gas source in Africa. The recoverable unconventional gas resources in Algeria are 20.7×10^{12} m^3, which account for 48.3% of the total recoverable unconventional gas resources in Africa and mainly consist of shale gas. The recoverable unconventional gas resources in the Republic of South Africa, Egypt,

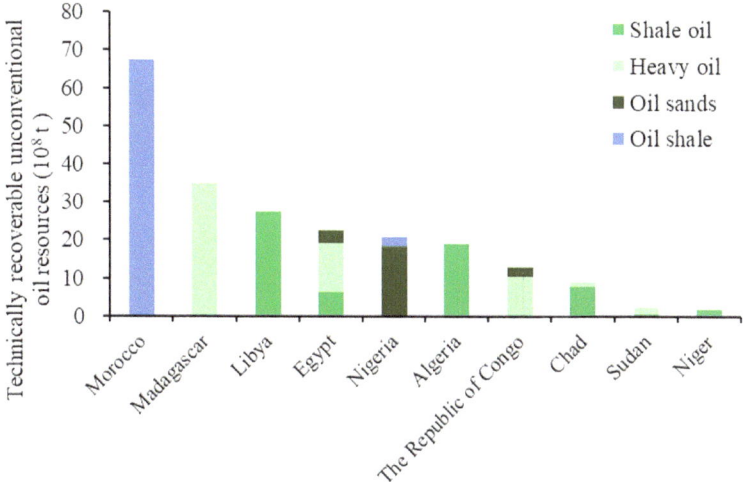

Fig. 7.20 Histogram of technically recoverable unconventional oil resources in major African countries (regions)

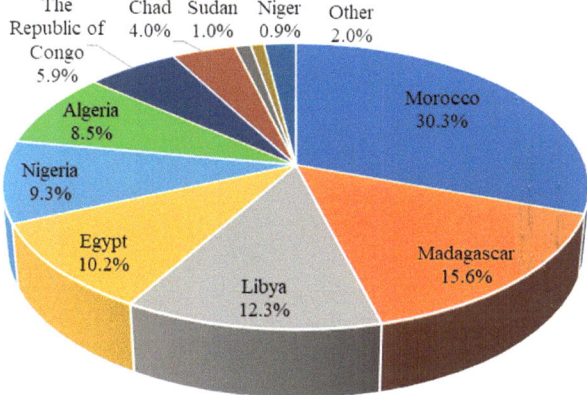

Fig. 7.21 Pie chart of technically recoverable unconventional oil resources in major African countries (regions)

and Libya mainly consist of shale gas and amount to 14.5×10^{12} m^3, 4.6×10^{12} m^3, and 2.9×10^{12} m^3, respectively, accounting for 33.8%, 10.8%, and 6.8% of the total recoverable unconventional gas resources in Africa (Figs. 7.22 and 7.23).

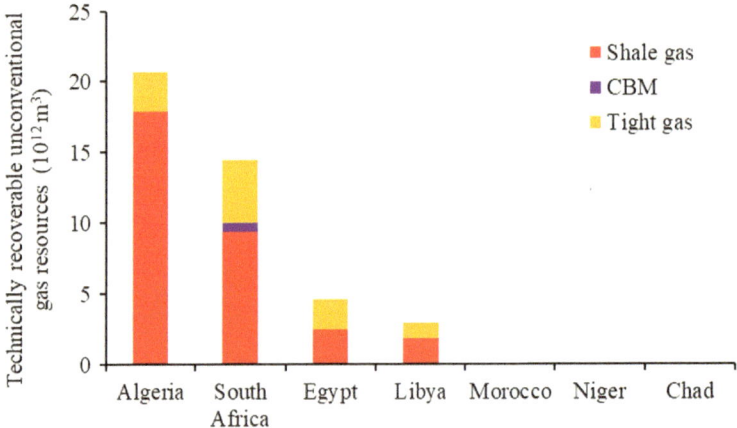

Fig. 7.22 Histogram of technically recoverable unconventional gas resources in major African countries (regions)

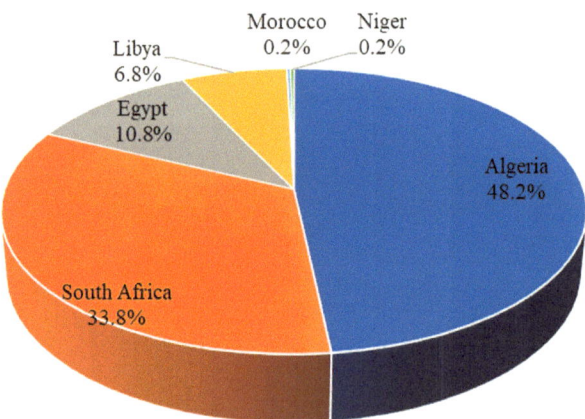

Fig. 7.23 Pie chart of technically recoverable unconventional gas resources in major African countries (regions)

7.2.2 Distribution of Unconventional Oil and Gas Resources in Major Basins

The recoverable unconventional oil and gas resources in Africa are mainly distributed in 25 basins, including the Ghadames Basin, Tindouf Basin, Sirte Basin, Northern Egypt Basin, Abu Garadi Basin, Bongor Basin, Eastern Niger Basin, Nile Delta Basin, and Illizi Basin. The recoverable unconventional oil resources are distributed in 23 basins, and the recoverable unconventional gas resources are distributed in 13 basins.

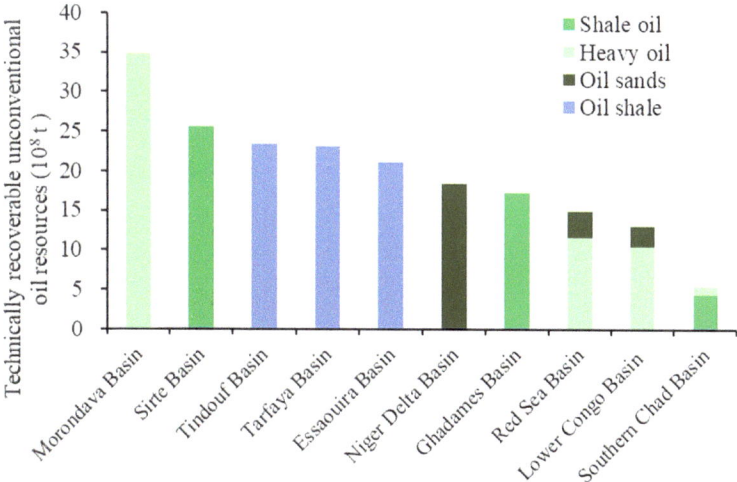

Fig. 7.24 Histogram of technically recoverable unconventional oil resources in major basins in Africa

7.2.2.1 Distribution of Unconventional Oil Resources in Major Basins

88.4% of the recoverable unconventional oil and gas resources in Africa are concentrated in the Morondava Basin, Sirte Basin, Tindouf Basin, Tarfaya Basin, Essaouira Basin, Niger Delta Basin, Ghadames Basin, Red Sea Basin, Lower Congo Basin, and Southern Chad Basin. The Morondava Basin ranks first in terms of recoverable unconventional oil resources, and the recoverable heavy oil resources in this basin amount to 34.8×10^8 tons, accounting for 15.6% of the total recoverable unconventional oil resources in Africa. The Sirte Basin ranks second, and the recoverable shale oil resources in this basin are 25.6×10^8 tons, accounting for 11.5%. The Tindouf Basin ranks third, and the recoverable oil shale in this basin is 23.3×10^8 tons, accounting for 10.5%.

The oil shale in Africa is concentrated in three basins, namely, the Tindouf Basin, the Tarfaya Basin, and the Essaouira Basin. The heavy oil resources in Africa are mainly distributed in the Morondava Basin, Red Sea Basin, and Lower Congo Basin; the shale oil resources in Africa are mainly distributed in the Sirte Basin, Ghadames Basin, and Southern Chad Basin; and the oil sands in Africa are concentrated in the Niger Delta Basin (Figs. 7.24 and 7.25).

7.2.2.2 Distribution of Unconventional Gas Resources in Major Basins

92.6% of Africa's unconventional gas resources, most of which are shale gas, are concentrated in the Ghadames Basin, Karoo Basin, Ahnet Basin, Regan Basin, Sirte Basin, Abu Garadi Basin, and Illizi Basin. The Ghadames Basin ranks first in Africa

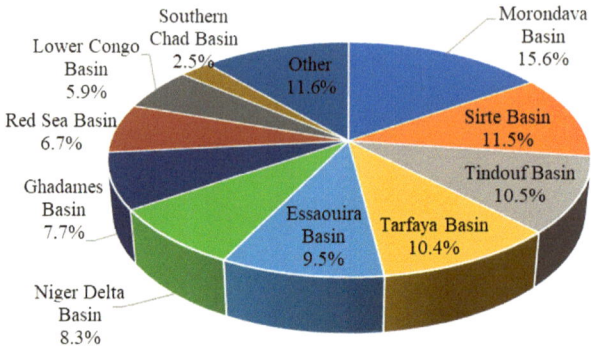

Fig. 7.25 Pie chart of technically recoverable unconventional oil resources in major basins in Africa

in terms of recoverable unconventional gas resources, and the recoverable shale gas resources in this basin are 10.9×10^{12} m^3, accounting for 25.4% of the total recoverable unconventional gas resources in Africa. The Karoo Basin in Algeria ranks second, and the recoverable shale gas resources in this basin are 9.1×10^{12} m^3, accounting for 21.2% of the total recoverable unconventional gas resources in Africa. The Ahnet Basin ranks third, and the recoverable shale gas resources in this basin are 6.6×10^{12} m^3, accounting for 15.3% of the total recoverable unconventional gas resources in Africa (Figs. 7.26 and 7.27).

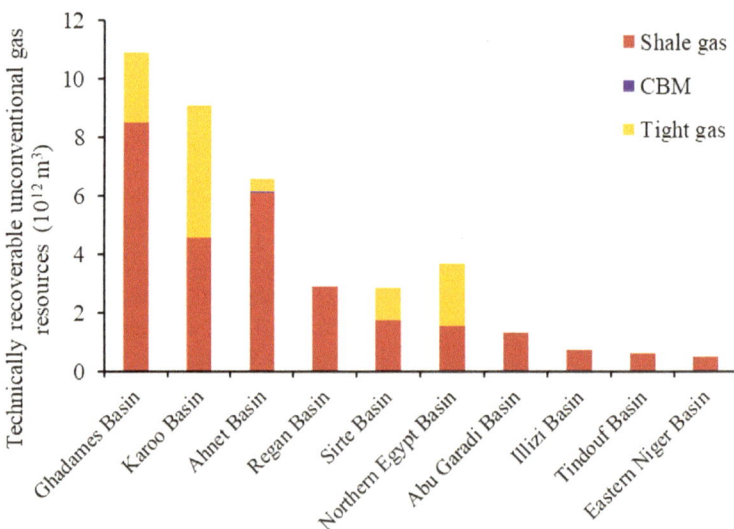

Fig. 7.26 Histogram of technically recoverable unconventional gas resources in major basins in Africa

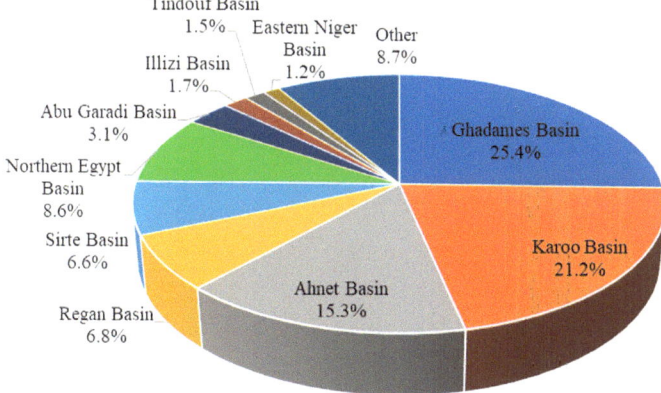

Fig. 7.27 Pie chart of technically recoverable unconventional gas resources in major basins in Africa

Chapter 8
Distribution of Oil and Gas Resources in the Middle East

The Middle East includes 24 countries and regions, including Bahrain, Iran, Iraq, Israel, Jordan, Kuwait, Lebanon, Oman, Qatar, Saudi Arabia, Syria, the United Arab Emirates (UAE), Yemen, Palestine, and Turkey. Nine major sedimentary basins have developed in the Middle East, and cover a total area of about 626×10^4 km². The main types of onshore basins are foreland basins and passive margin basins, and the main types of offshore basins are rift basins and passive margin basins. The Middle East has the most abundant oil and gas resources, and 22.1% of the world's oil and gas resources are concentrated in this region. The recoverable oil and gas resources in the Middle East amount to 4124.4×10^8 tons of oil equivalent.

8.1 Conventional Oil and Gas Resources

The Middle East has abundant oil and gas resources. The conventional oil and gas resources in the Middle East amount to 3612.9×10^8 tons of oil equivalent, accounting for 31.6% of the total conventional oil and gas resources worldwide. The recoverable conventional oil and gas resources in the Middle East are 2535.2×10^8 tons of oil equivalent, accounting for 38.9% of the world's total recoverable conventional resources. The remaining recoverable reserves of oil and gas in the Middle East amount to 1979.6×10^8 tons of oil equivalent, accounting for 50.7% of the world's total remaining recoverable reserves. The predicted reserve growth from discovered oil and gas fields in the Middle East are 397.4×10^8 tons, accounting for 35.5% of the world's total reserve growth in the future. The undiscovered recoverable oil and gas resources the Middle East are 680.4×10^8 tons of oil equivalent, accounting for 17.9% of the total undiscovered recoverable resources worldwide.

© The Author(s) 2024
L. Dou et al., *Global Oil and Gas Resources: Potential and Distribution*,
https://doi.org/10.1007/978-981-97-4756-6_8

8.1.1 Distribution of Remaining Recoverable Reserves

The remaining recoverable reserves of oil and gas in the Middle East are about 1979.6 $\times$ 10^8 tons of oil equivalent, including 911.8 $\times$ 10^8 tons of oil, 105.9 $\times$ 10^8 tons of condensate, and 112.5 $\times$ 10^{12} m^3 of gas.

8.1.1.1 Distribution of Remaining Recoverable Reserves in Major Countries

Qatar has the largest amounts of remaining recoverable reserves. The remaining recoverable reserves of oil and gas in Qatar are 494.8 $\times$ 10^8 tons of oil equivalent, of which oil accounts for 1.1%, condensate accounts for 11.8%, and gas accounts for 87.1%. It is obvious that, in this country, the remaining recoverable gas reserves are significantly more than the remaining recoverable oil reserves. The remaining recoverable oil and gas reserves in Saudi Arabia and Iran rank second and third, respectively. In these two countries, the remaining recoverable reserves of gas are more than oil (Figs. 8.1 and 8.2). The remaining recoverable reserves in Saudi Arabia are about 488.7 $\times$ 10^8 tons of oil equivalent, of which oil accounts for 78.3%, condensate accounts for 2.0%, and gas accounts for 19.7%. The remaining recoverable reserves in Iran are 443.1 $\times$ 10^8 tons of oil equivalent, of which oil accounts for 32.5%, condensate accounts for 5.6%, and gas accounts for 61.9%.

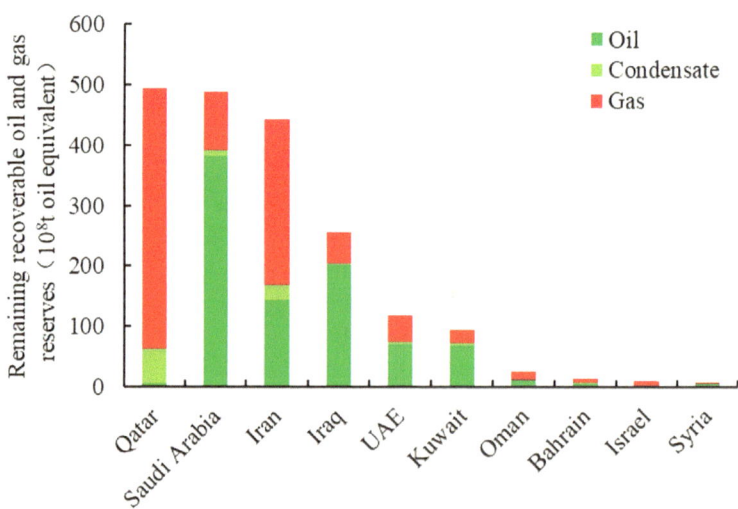

Fig. 8.1 Histogram of remaining recoverable reserves in major countries (regions) in the Middle East

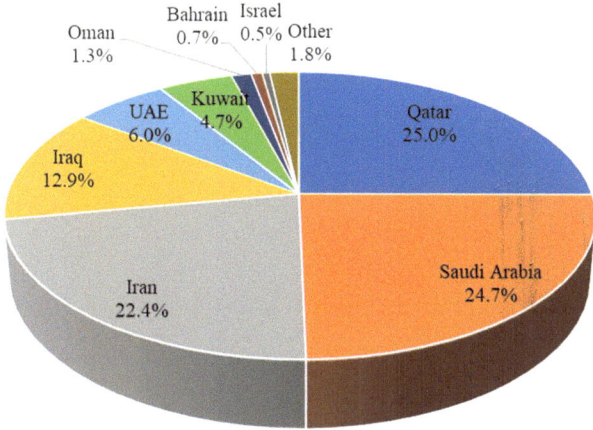

Fig. 8.2 Pie chart of remaining recoverable reserves in major countries (regions) in the Middle East

8.1.1.2 Distribution of Remaining Recoverable Reserves in Major Basins

The remaining recoverable oil and gas reserves in the Middle East are mainly concentrated in the Arabian Basin, the Zagros Basin, the Oman Basin, the Levant Basin, and Black Sea Basin. The Arabian Basin, Zagros Basin, and Oman Basin rank among the top three in terms of remaining recoverable reserves. The remaining recoverable oil and gas reserves in these three basins account for 98.5% of the total remaining recoverable reserves in the Middle East (Figs. 8.3 and 8.4).

The remaining recoverable reserves in the Arabian Basin are 1615.9×10^8 tons of oil equivalent, of which oil accounts for 46.0%, condensate accounts for 5.6%, and gas accounts for 48.3%. The remaining recoverable reserves in the Zagros Basin are 306.7×10^8 tons of oil equivalent, of which oil accounts for 50.1%, condensate accounts for 3.8%, and gas accounts for 46.1%. The remaining recoverable reserves in the Oman Basin are 27.7×10^8 tons of oil equivalent, of which oil accounts for 41.4%, condensate accounts for 7.6%, and gas accounts for 50.9%.

8.1.1.3 Distribution of Remaining Recoverable Reserves in Onshore and Offshore Areas

The remaining recoverable reserves of oil and gas in offshore and onshore areas of the Middle East account for 39.8% and 60.2%, respectively (Fig. 8.5). The remaining recoverable reserves in offshore areas are 972.0×10^8 tons of oil equivalent, of which oil accounts for 27.1%, condensate accounts for 5.1%, and gas accounts for 67.8%. The remaining recoverable reserves in onshore areas are 643.9×10^8 tons of oil equivalent, of which oil accounts for 65.2%, condensate accounts for 3.7%, and gas accounts for 31.1%.

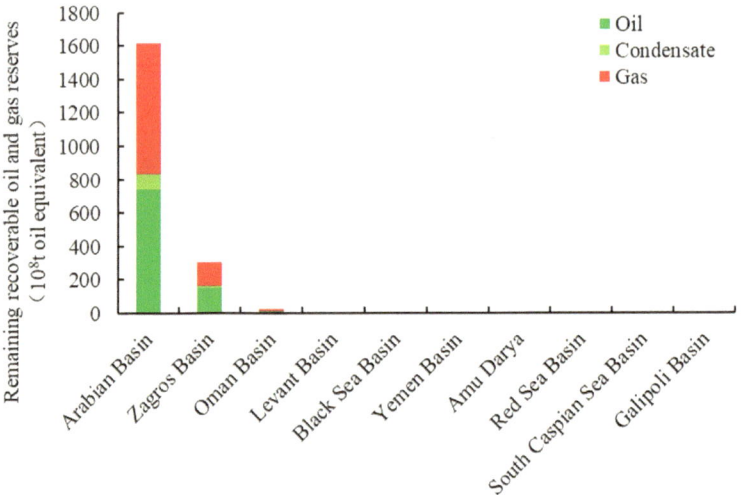

Fig. 8.3 Histogram of remaining recoverable reserves in major basins in the Middle East

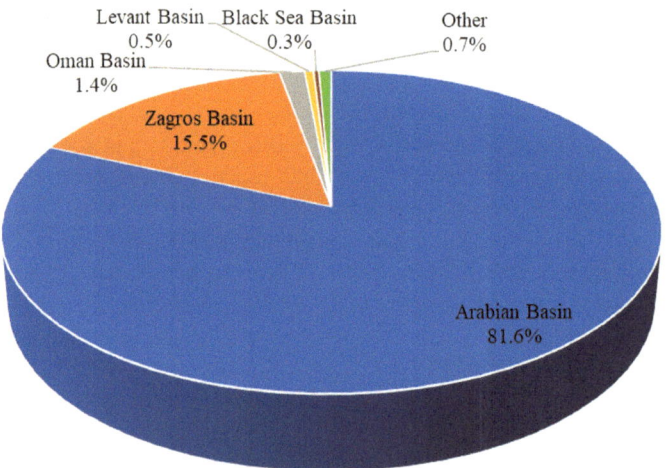

Fig. 8.4 Pie chart of remaining recoverable reserves in major basins in the Middle East

8.1.2 *Trend of Reserve Growth from Discovered Oil and Gas Fields*

The future reserve growth from discovered oil and gas fields in the Middle East is nearly 397.4×10^8 tons of oil equivalent, including of 167.1×10^8 tons oil, 34.7×10^8 tons of condensate, and 23.0×10^{12} m^3 of gas.

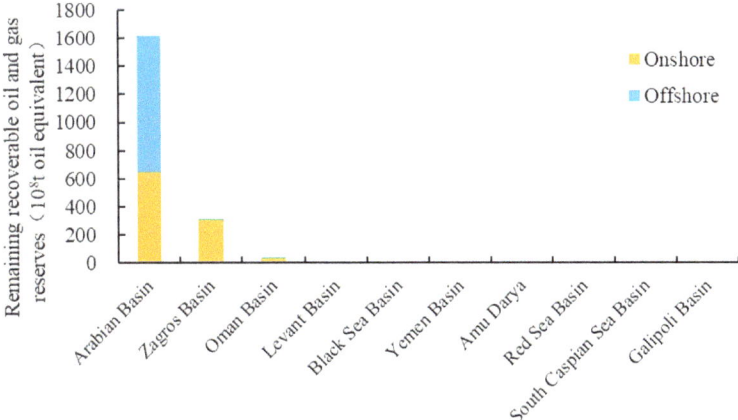

Fig. 8.5 Histogram of remaining recoverable reserves in onshore and offshore areas of major basins in the Middle East

8.1.2.1 Distribution of Reserve Growth in Major Countries (Regions)

The discovered oil and gas fields in Saudi Arabia contribute the most to future reserve growth in the Middle East. The future reserve growth from discovered oil and gas fields in Saudi Arabia is nearly 102.9×10^8 tons of oil equivalent, accounting for 25.8% of the total reserve growth in the Middle East, of which oil accounts for 42.6%, condensate accounts for 9.9%, and gas accounts for 47.5%. The reserve growth from discovered oil and gas fields in Qatar ranks second and amounts to about 77.3×10^8 tons of oil equivalent, of which oil accounts for 9.0%, condensate accounts for 9.9%, and gas accounts for 81.1%. The reserve growth from discovered oil and gas fields in Iran ranks third and amounts to 74.2×10^8 tons of oil equivalent, of which gas and oil account for 47.7% and 45.8%, respectively (Figs. 8.6 and 8.7).

8.1.2.2 Distribution of Reserve Growth in Major Basins

The future reserve growth in discovered oil and gas fields in the Middle East are mainly concentrated in the Arabian Basin, the Zagros Basin, the Levant Basin, the Oman Basin, and Yemen Basin. The reserve growth in the Arabian Basin, Zagros Basin and Levant Basin, which are the top three basins with the largest amounts of additional oil and gas reserves in the Middle East, account for about 98.0% of the total reserve growth from discovered oil and gas fields in the Middle East (Figs. 8.8 and 8.9).

The future reserve growth from discovered oil and gas fields in the Arabian Basin is 286.4×10^8 tons of oil equivalent, of which oil accounts for 42.6%, condensate accounts for 9.9%, and gas accounts for 47.5%. The future reserve growth from discovered oil and gas fields in the Zagros Basin are 92.7×10^8 tons of oil equivalent,

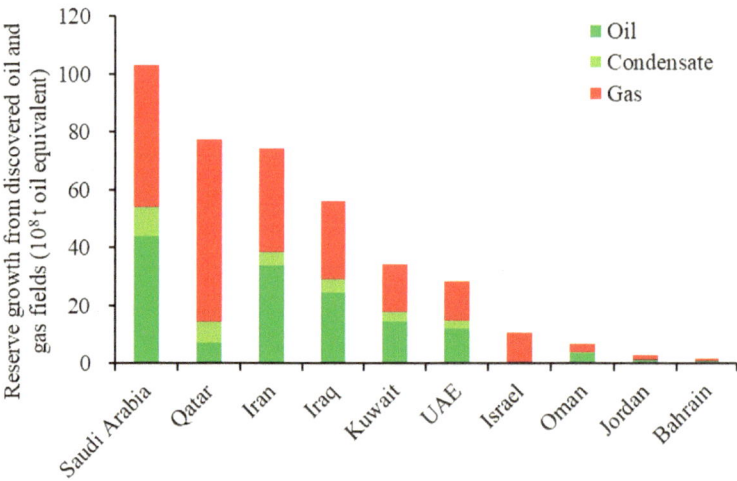

Fig. 8.6 Histogram of future reserve growth in discovered oil and gas fields in major countries (regions) in the Middle East

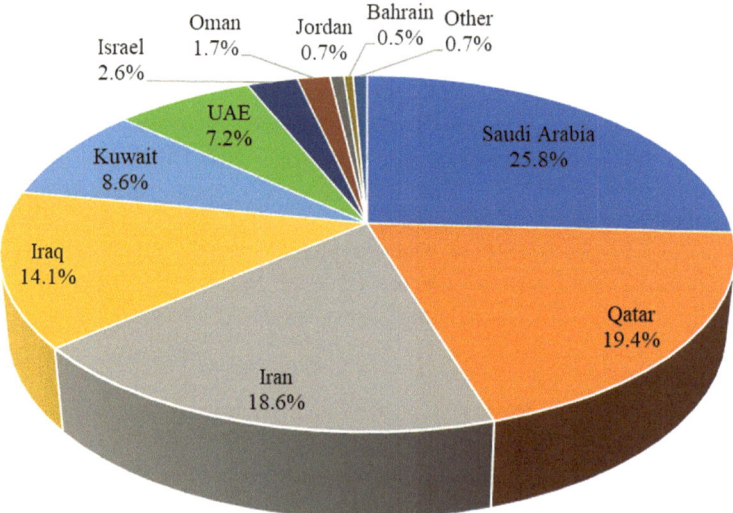

Fig. 8.7 Pie chart of future reserve growth in discovered oil and gas fields in major countries (regions) in the Middle East

of which oil accounts for 45.8%, condensate accounts for 6.5%, and gas accounts for 47.7%. The future reserve growth from discovered oil and gas fields in the Levant Basin is 10.5×10^8 tons of oil equivalent, of which oil accounts for 0.2%, condensate accounts for 0.5%, and gas accounts for 99.3%.

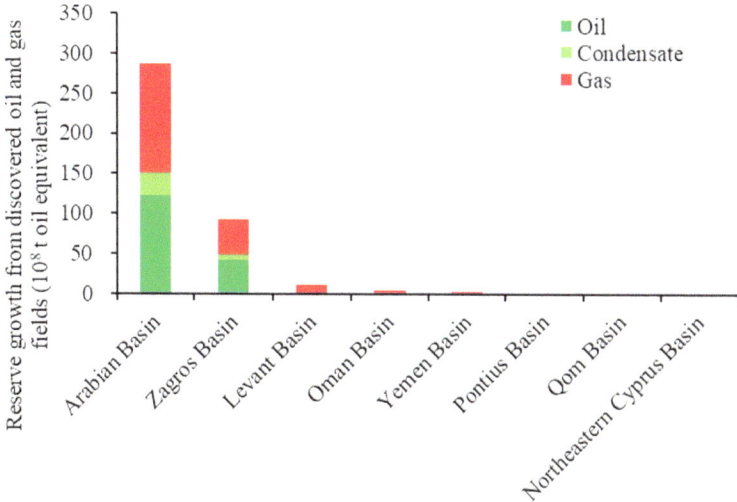

Fig. 8.8 Histogram of future reserve growth in discovered oil and gas fields in major basins in the Middle East

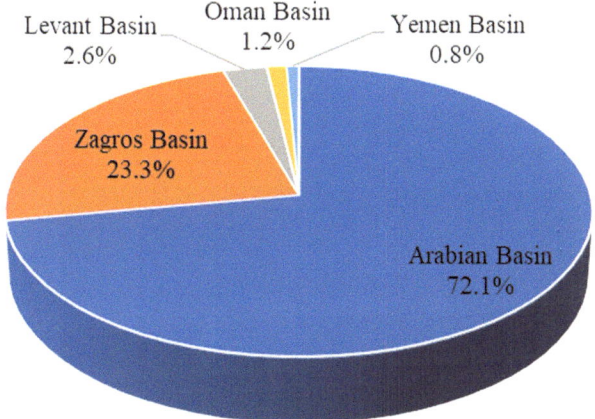

Fig. 8.9 Pie chart of future reserve growth in discovered oil and gas fields in major basins in the Middle East

8.1.2.3 Distribution of Reserve Growth in Onshore and Offshore Areas

The reserve growth in the Middle East is unevenly distributed in onshore and offshore oil and gas fields. The reserve growth in onshore oil and gas fields is 237.1×10^8 tons of oil equivalent, of which oil accounts for 72.2%, condensate accounts for 2.8%, and gas accounts for 25.0%. The reserve growth in offshore oil and gas fields is 160.3×10^8 tons of oil equivalent, of which oil accounts for 32.1%, condensate accounts for 4.8%, and gas accounts for 63.1% (Fig. 8.10).

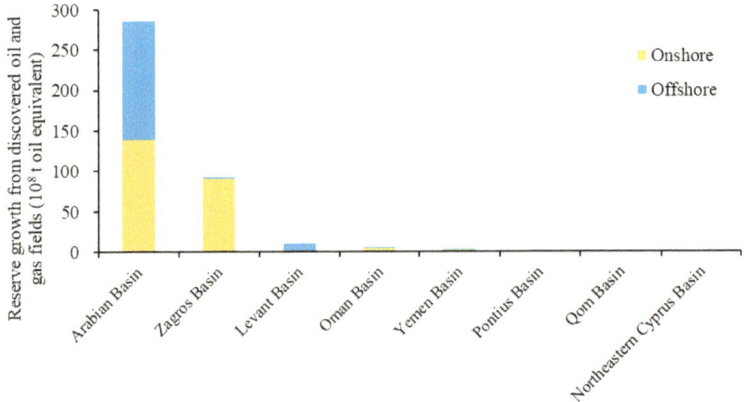

Fig. 8.10 Histogram of future reserve growth in onshore and offshore oil and gas fields in major basins in the Middle East

8.1.3   Characteristics of Distributions of Undiscovered Oil and Gas Resources

The undiscovered oil and gas resources in the Middle East amount to 680.4 × 10^8 tons, including 296.0 × 10^8 tons of oil, 47.5 × 10^8 tons of condensate, and 39.4 × 10^{12} m^3 of gas.

8.1.3.1   Distribution of Undiscovered Oil and Gas Resources in Major Countries (Regions)

Iran has the largest amounts of undiscovered oil and gas resources, which are more than 213.0 × 10^8 tons of oil equivalent and mainly consist of gas. Among the undiscovered oil and gas resources in Iran, oil accounts for 35.1%, condensate accounts for 5.9%, and gas accounts for 59.0%. The undiscovered oil and gas resources in Saudi Arabia amount to 124.6 × 10^8 tons of oil equivalent, of which oil accounts for 49.3%, condensate accounts for 7.2%, and gas accounts for 43.5%. The undiscovered oil and gas resources in Iraq are 106.8 × 10^8 tons of oil equivalent, of which oil accounts for 46.2%, condensate accounts for 6.8%, and gas accounts for 47.0% (Figs. 8.11 and 8.12).

8.1.3.2   Distribution of Undiscovered Oil and Gas Resources in Major Basins

The undiscovered recoverable oil and gas resources in the Middle East are mainly distributed in the Arabian Basin, Zagros Basin, Levant Basin, Oman Basin, Qom

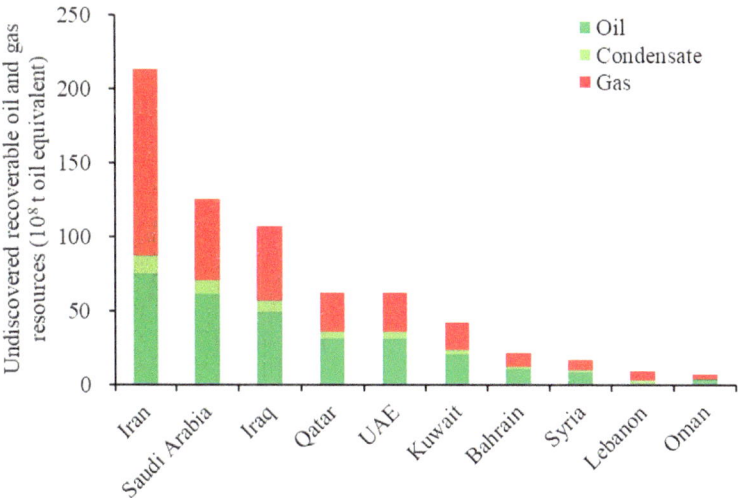

Fig. 8.11 Histogram of undiscovered recoverable oil and gas resources in major countries (regions) in the Middle East

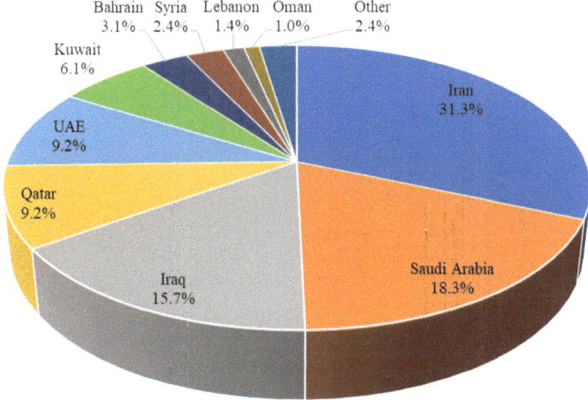

Fig. 8.12 Pie chart of undiscovered recoverable oil and gas resources in major countries (regions) in the Middle East

Basin, and Yemen Basin. The Arabian Basin, Zagros Basin and Levant Basin rank among the top three in terms of undiscovered oil and gas resources. The sum of undiscovered recoverable oil and gas resources in these three basins accounts for 98.5% of the total undiscovered recoverable resources in the Middle East (Figs. 8.13 and 8.14).

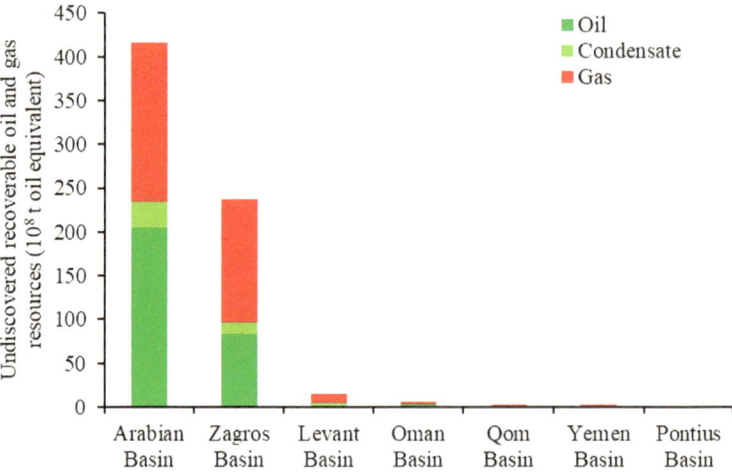

Fig. 8.13 Histogram of undiscovered recoverable oil and gas resources in major basins in the Middle East

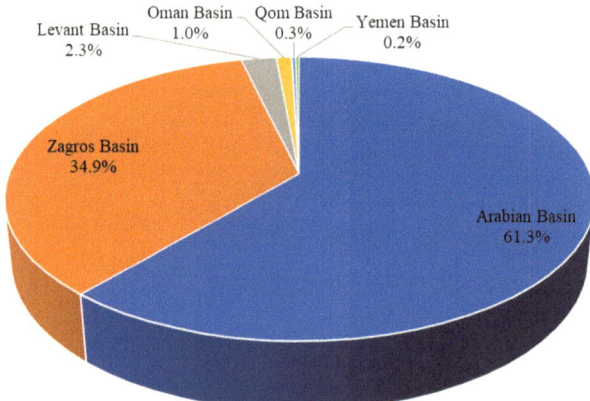

Fig. 8.14 Pie chart of undiscovered recoverable oil and gas resources in major basins in the Middle East

The undiscovered oil and gas resources in the Arabian Basin amount to 415.4×10^8 tons of oil equivalent, of which oil accounts for 49.3%, condensate accounts for 7.2%, and gas accounts for 43.5%. The undiscovered oil and gas resources in the Zagros Basin are 236.7×10^8 tons of oil equivalent, of which oil accounts for 35.1%, condensate accounts for 5.9%, and gas accounts for 59.0%. The undiscovered oil and gas resources in the Levant Basin are 15.4×10^8 tons of oil equivalent, which are mainly located in deep to ultra-deep waters and of which oil accounts for 10.7%, condensate accounts for 23.5%, and gas accounts for 65.8%.

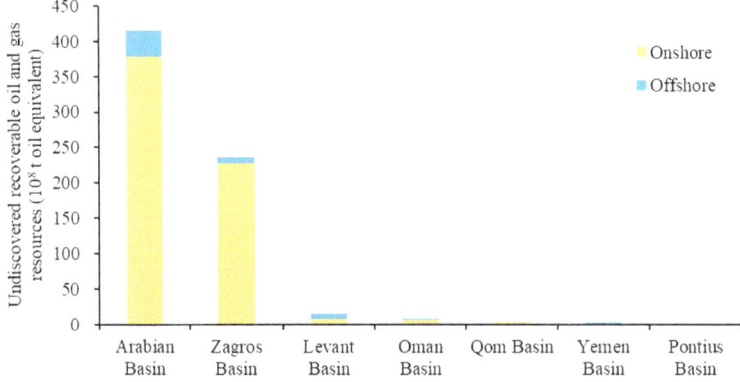

Fig. 8.15 Histogram of undiscovered recoverable oil and gas resources in onshore and offshore areas of major basins in the Middle East

8.1.3.3 Distribution of Undiscovered Oil and Gas Resources in Onshore and Offshore Areas

In the Middle East, the undiscovered onshore oil and gas resources are significantly more than the undiscovered offshore oil and gas resources (Fig. 8.15). The undiscovered onshore oil and gas resources in the Middle East amount to 621.6×10^8 tons of oil equivalent, of which oil accounts for 43.8%, condensate accounts for 6.8%, and gas accounts for 49.4%. In contrast, the undiscovered offshore oil and gas resources in the Middle East are only 58.8×10^8 tons of oil equivalent, of which oil accounts for 41.4%, condensate accounts for 9.4%, and gas accounts for 49.2%.

8.2 Distribution of Unconventional Oil and Gas Resources

The unconventional oil and gas resources in the Middle East include oil shale, heavy oil, shale oil, shale gas, and tight gas. The technically recoverable unconventional oil and gas resources in the Middle East are 511.5×10^8 tons of oil equivalent, accounting for 7.1% of the world's total recoverable unconventional resources.

The technically recoverable unconventional oil resources in the Middle East are 302.2×10^8 tons of oil equivalent, accounting for 7.1% of the world's total recoverable unconventional oil resources and including about 180.6×10^8 tons of heavy oil, 62.6×10^8 tons of oil shale, and 59.0×10^8 tons of shale oil (Figs. 8.16 and 8.17).

The technically recoverable unconventional gas resources in the Middle East are 24.5×10^{12} m^3, including 16.1×10^{12} m^3 of shale gas and 8.4×10^{12} m^3 of tight gas (Figs. 8.18 and 8.19).

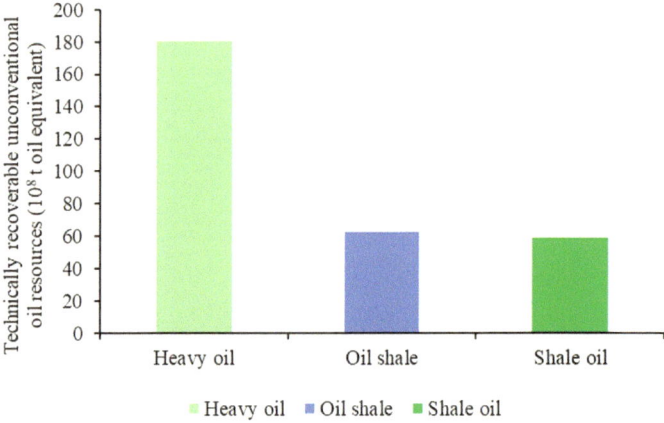

Fig. 8.16 Histogram of technically recoverable unconventional oil resources in the Middle East

Fig. 8.17 Pie chart of technically recoverable unconventional oil resources in the Middle East

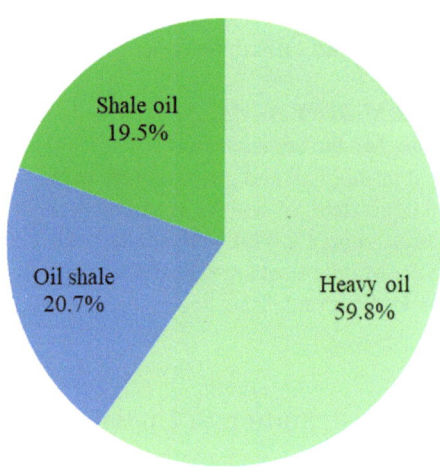

8.2.1   Distribution of Recoverable Unconventional Oil and Gas Resources in Major Countries (Regions)

8.2.1.1   Distribution of Recoverable Unconventional Oil Resources in Major Countries (Regions)

The unconventional oil resources in the Middle East are mainly distributed in Saudi Arabia, the UAE, Iraq, Iran, and Oman. The unconventional oil resources in Saudi Arabia are 226.0×10^8 tons of oil equivalent, including 160.2×10^8 tons of heavy oil, 57.0×10^8 tons of oil shale, and 8.8×10^8 tons of shale oil. The recoverable unconventional oil resources in the UAE are 33.9×10^8 tons, all of which are shale oil.

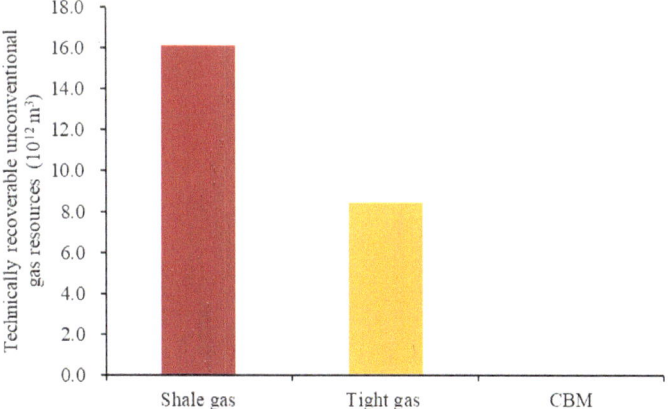

Fig. 8.18 Histogram of technically recoverable unconventional gas resources in the Middle East

Fig. 8.19 Pie chart of technically recoverable unconventional gas resources in the Middle East

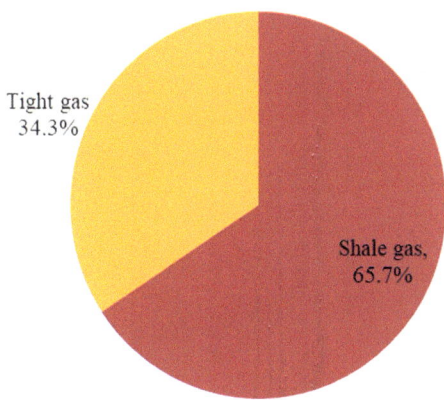

The recoverable unconventional oil resources in Iraq are 23.6×10^8 tons, including 20.4×10^8 tons of heavy oil and 3.2×10^8 tons of shale oil (Figs. 8.20 and 8.21).

8.2.1.2 Distribution of Recoverable Unconventional Gas Resources in Major Countries (Regions)

The unconventional gas resources in the Middle East are mainly distributed in Saudi Arabia, the UAE, Oman and Jordan. The recoverable unconventional gas resources in Saudi Arabia are 11.9×10^{12} m^3, including 8.7×10^{12} m^3 of shale gas and 3.2×10^{12} m^3 of tight gas. The recoverable unconventional gas resources in the UAE are 6.5×10^{12} m^3, all of which are shale gas (Figs. 8.22 and 8.23).

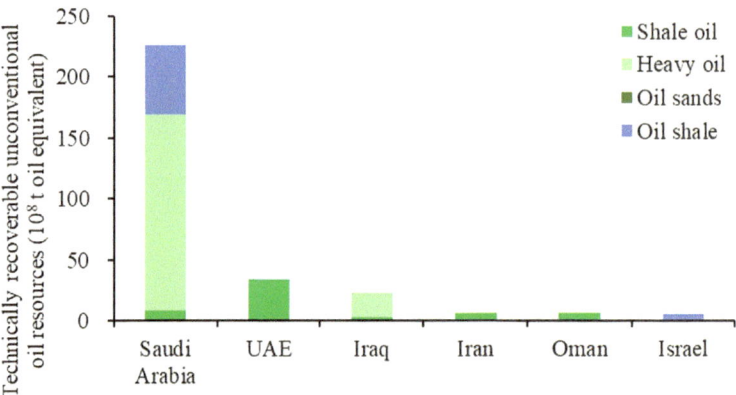

Fig. 8.20 Histogram of technically recoverable unconventional oil resources in major countries in the Middle East

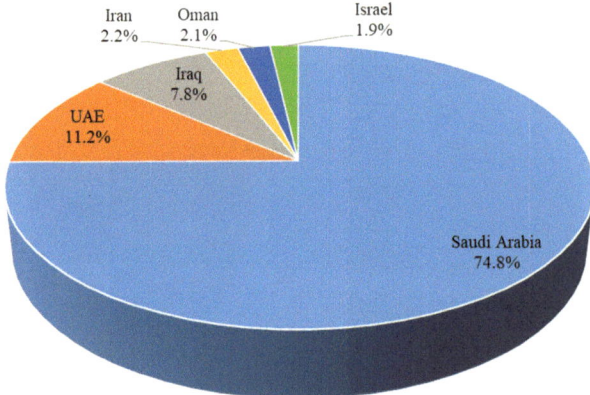

Fig. 8.21 Pie chart of technically recoverable unconventional oil resources in major countries in the Middle East

8.2.2   Distribution of Unconventional Oil and Gas Resources in Major Basins

8.2.2.1   Distribution of Recoverable Unconventional Oil Resources in Major Basins

The unconventional oil resources in the Middle East are mainly concentrated in the Arabian Basin, Zagros Basin, and Oman Basin (Figs. 8.24 and 8.25).

In the Middle East, shale oil resources are mainly distributed in Arabian Basin, Zagros Basin, and Oman Basin; heavy oil resources are mainly distributed in the Arabian Basin and Zagros Basin; and oil shale are mainly distributed in the Arabian

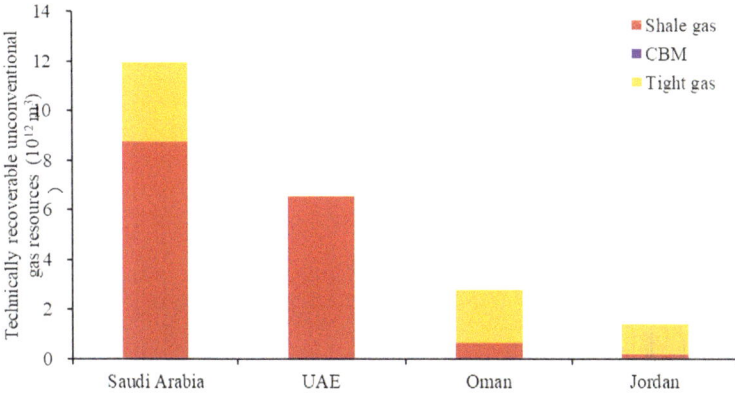

Fig. 8.22 Histogram of technically recoverable unconventional gas resources in major countries in the Middle East

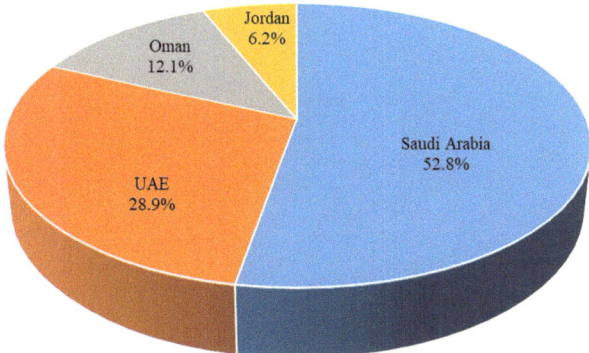

Fig. 8.23 Pie chart of technically recoverable unconventional gas resources in major countries in the Middle East

Basin and Levant Basin. The Arabian Basin ranks first in the Middle East in terms of recoverable unconventional oil resources. The recoverable unconventional oil resources in the Arabian Basin amount to 260×10^8 tons, of which heavy oil accounts for 61.6%, oil shale accounts for 21.9%, and shale oil accounts for 16.5%. The recoverable unconventional oil resources in the Zagros Basin are 30.1×10^8 tons, which mainly consist of shale oil and heavy oil and of which shale oil accounts for 32.4% and heavy oil accounts for 67.6%.

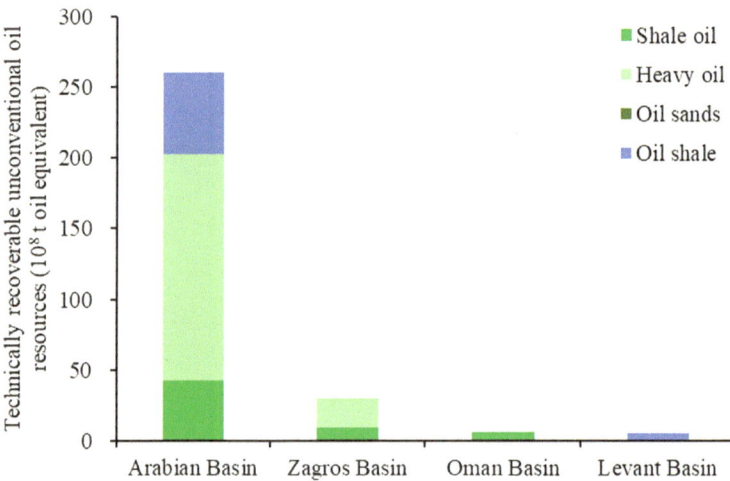

Fig. 8.24 Histogram of technically recoverable unconventional oil resources in major basins in the Middle East

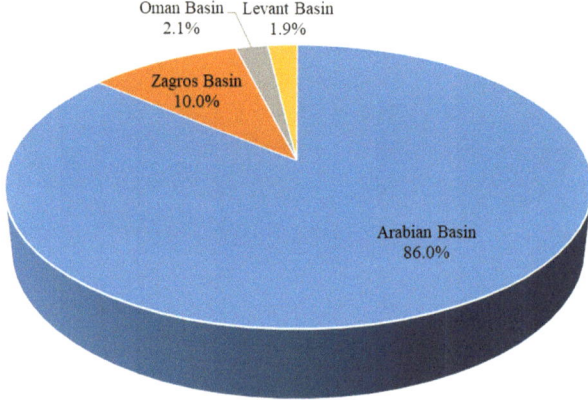

Fig. 8.25 Pie chart of technically recoverable unconventional oil resources in major basins in the Middle East

8.2.2.2 Distribution of Recoverable Unconventional Gas Resources in Major Basins

The recoverable unconventional gas resources in the Middle East are mainly concentrated in the Arabian Basin, Oman Basin, and Zagros Basin (Figs. 8.26 and 8.27). The Arabian Basin has the most abundant recoverable unconventional gas resources, which amount 19.5×10^{12} m^3, accounting for more than 80.1% of the total recoverable unconventional gas resources in all basins in the Middle East and mainly consisting of shale gas and tight gas. The recoverable shale gas resources and tight

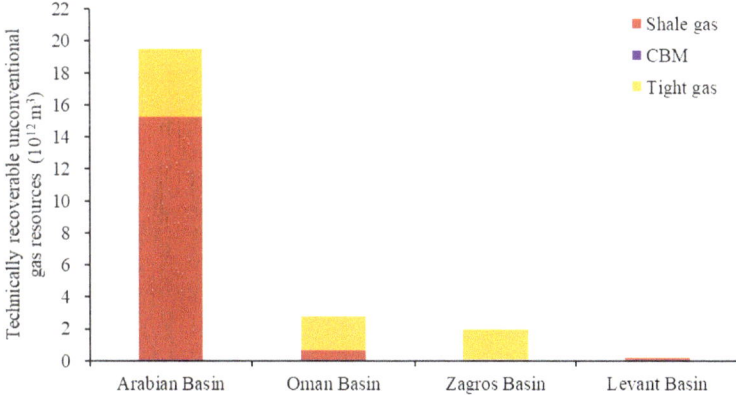

Fig. 8.26 Histogram of technically recoverable unconventional gas resources in major basins in the Middle East

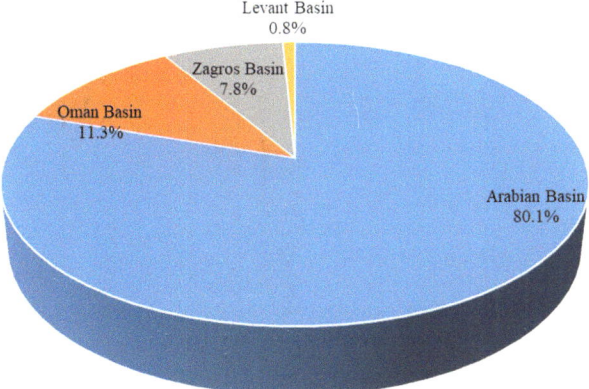

Fig. 8.27 Pie chart of technically recoverable unconventional gas resources in major basins in the Middle East

gas resources in the Arabian Basin account for 98.8% and 50% of the unconventional gas resources in the basin, respectively. The Oman Basin ranks second in terms of unconventional gas resources. The recoverable unconventional gas resources in this basin are 2.7×10^{12} m^3, including shale gas and tight gas.

Chapter 9
Distribution of Oil and Gas Resources in Central Asia

The Central Asia includes Kazakhstan, Kyrgyzstan, Uzbekistan, Tajikistan, Turkmenistan, Azerbaijan, Georgia and other countries. Eighteen oil and gas-bearing basins have developed in Central Asia, covering a total area of 416.29×10^4 km^2. The total area of onshore sedimentary basins in Central Asia is about 277.0×10^4 km^2, and that of offshore sedimentary basins is about 33.2×10^4 km^2. About 4.9% of the world's oil and gas resources are distributed across Central Asia. The total recoverable oil and gas resources in Central Asia amount to 924.3×10^8 tons of oil equivalent.

9.1 Conventional Oil and Gas Resources

The recoverable conventional oil and gas resources in Central Asia are 716.7×10^8 tons of oil equivalent, accounting for 6.3% of the world's total recoverable conventional resources. The recoverable oil and gas reserves in Central Asia are 337.5×10^8 tons, accounting for 5.2% of the world's total recoverable reserves. The cumulative oil and gas production in Central Asia are 133.8×10^8 tons of oil equivalent, accounting for 5.1% of the world's total cumulative oil and gas production. The remaining recoverable reserves of oil and gas in Central Asia are 203.7×10^8 tons of oil equivalent, accounting for 5.2% of the world's total remaining recoverable reserves. The predicted reserve growth from discovered oil and gas fields in Central Asia are 96.0×10^8 tons of oil equivalent, accounting for 8.6% of the world's total estimated reserve growth. The undiscovered recoverable oil and gas resources in Central Asia are 283.2×10^8 tons of oil equivalent, accounting for 7.5% of the world's total undiscovered recoverable resources.

© The Author(s) 2024
L. Dou et al., *Global Oil and Gas Resources: Potential and Distribution*,
https://doi.org/10.1007/978-981-97-4756-6_9

9.1.1 Distribution of Remaining Recoverable Reserves

The remaining recoverable reserves of oil and gas in Central Asia are 203.7 $\times$ 10^8 tons of oil equivalent, including 47.1 $\times$ 10^8 tons of oil, accounting for 23.1%, 8.1 $\times$ 10^8 tons of condensate, accounting for 4.0%, and 17.4 $\times$ 10^{12} m^3 of gas, accounting for 72.9%.

9.1.1.1 Distribution of Remaining Recoverable Reserves in Major Countries (Regions)

About 90.4% of the remaining recoverable reserves in Central Asia are concentrated in three countries, namely, Turkmenistan, Kazakhstan, and Azerbaijan. The remaining recoverable reserves of oil and gas in Turkmenistan account for 48.4% of the total remaining recoverable reserves in Central Asia and mainly consist of gas. The remaining recoverable reserves of oil and gas in Kazakhstan rank second, accounting for 31.1% of the total remaining recoverable reserves in Central Asia. In Kazakhstan, the remaining recoverable gas reserves are slightly more than the remaining recoverable oil reserves. The remaining recoverable oil reserves in Kazakhstan rank first in Central Asia, and the remaining recoverable gas reserves in this country rank second in Central Asia. The remaining recoverable reserves of oil and gas in Azerbaijan rank third and account for 10.9% of the total remaining recoverable reserves in Central Asia. In Azerbaijan, the remaining recoverable gas reserves are more than the remaining recoverable oil reserves. The remaining recoverable reserves in Central Asia are also distributed in other countries, such as Uzbekistan, Tajikistan, Georgia, and Kyrgyzstan (Figs. 9.1 and 9.2).

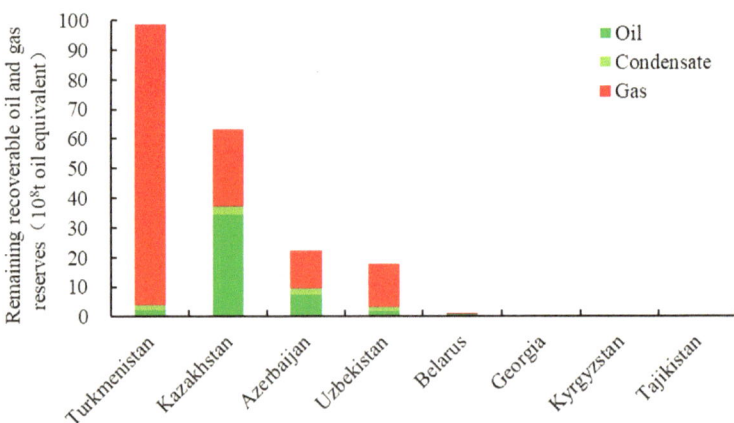

Fig. 9.1 Histogram of remaining recoverable reserves in major Central Asian countries

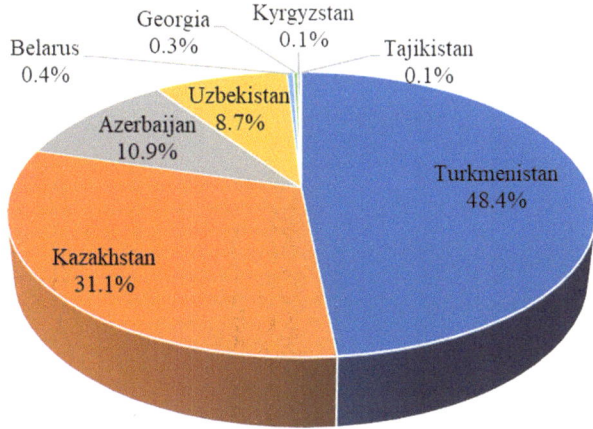

Fig. 9.2 Pie chart of remaining recoverable reserves in major Central Asian countries

9.1.1.2 Distribution of Remaining Recoverable Reserves in Major Basins

The remaining recoverable reserves of oil and gas in Central Asia are mainly distributed in the top ten basins (Fig. 9.3). The Amu Darya Basin, Precaspian Basin and South Caspian Basin are the top three basins in terms of remaining recoverable reserves. The sum of remaining recoverable reserves of oil and gas in these three basins accounts for 92.0% of the total remaining recoverable reserves in Central Asia (Fig. 9.4).

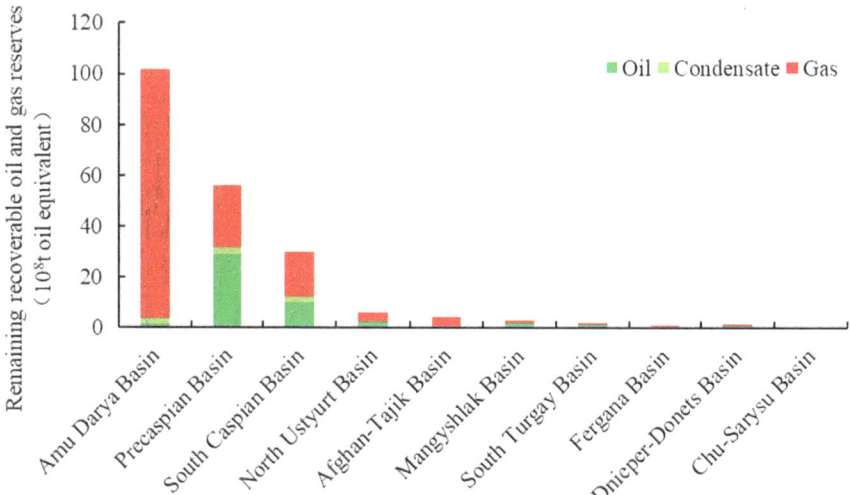

Fig. 9.3 Histogram of remaining recoverable reserves in top ten basins in Central Asia

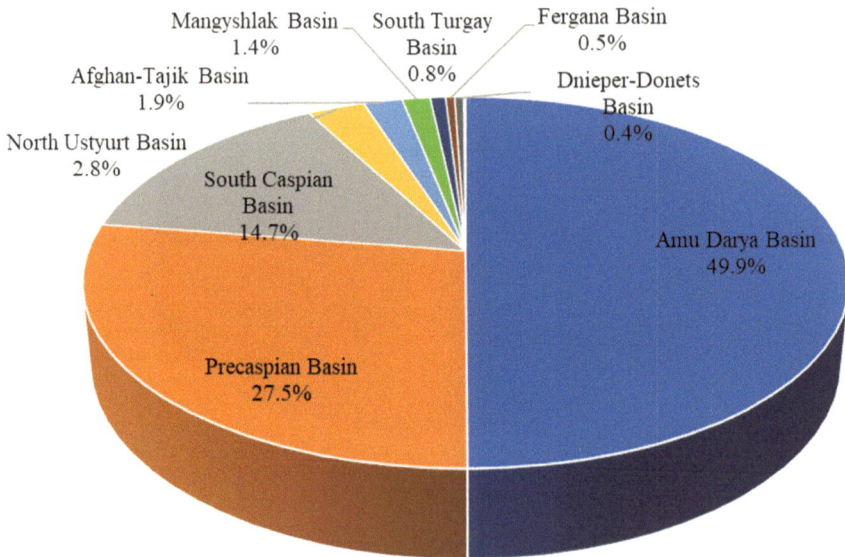

Fig. 9.4   Pie chart of remaining recoverable reserves distributed in major basins in Central Asia

The remaining recoverable reserves of oil and gas in the Amu Darya Basin amount to 101.6×10^8 tons of oil equivalent, of which oil accounts for 1.3%, condensate accounts for 2.4%, and gas accounts for 96.3%. The remaining recoverable reserves in the Precaspian Basin are 55.9×10^8 tons of oil equivalent, of which oil accounts for 52%, condensate accounts for 4.7%, and gas accounts for 43.3%. The remaining recoverable reserves in the South Caspian Basin are 29.9×10^8 tons of oil equivalent, of which oil accounts for 32.5%, condensate accounts for 8.3%, and gas accounts for 59.2%.

9.1.1.3   Distribution of Remaining Recoverable Reserves in Onshore and Offshore Areas

In Central Asia, the remaining recoverable reserves of oil and gas in onshore areas are much more than the remaining recoverable reserves in offshore areas. The remaining recoverable reserves in onshore and offshore areas account for 76.5% and 23.5% of the total remaining recoverable reserves in Central Asia, respectively (Fig. 9.5). The remaining recoverable reserves of onshore oil and gas amount to 155.9×10^8 tons of oil equivalent, of which oil accounts for 10.8%, condensate accounts for 5.1%, and gas accounts for 84.1%. The remaining recoverable reserves of offshore oil and gas are 47.8×10^8 tons of oil equivalent, of which oil accounts for 41.8%, condensate accounts for 5.0%, and gas accounts for 53.2%.

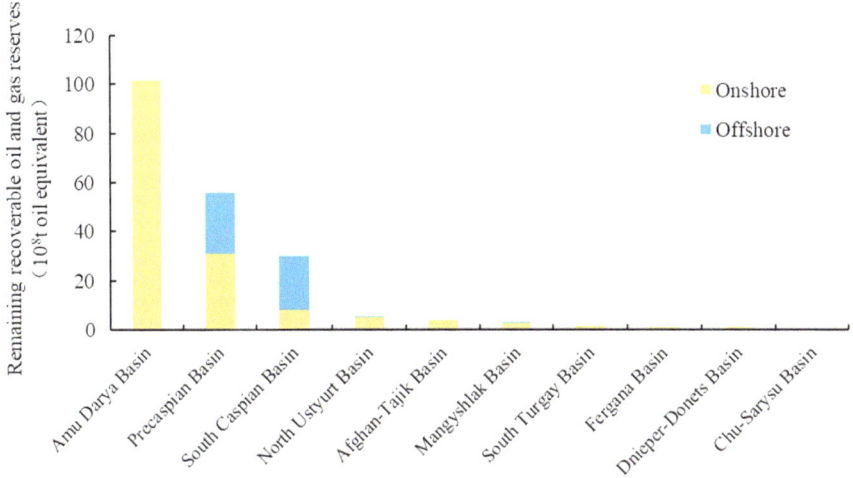

Fig. 9.5 Histogram of remaining recoverable reserves of onshore and offshore oil and gas in major basins in Central Asia

9.1.2 Trend of Reserve Growth from Discovered Oil and Gas Fields

The predicted reserve growth from discovered oil and gas fields in Central Asia are 96.0×10^8 tons of oil equivalent, including 20.6×10^8 tons of oil, accounting for 21.5%, 5.5×10^8 tons of condensate, accounting for 5.7%, and 8.2×10^{12} m^3 of gas, accounting for 72.8%.

9.1.2.1 Distribution of Reserve Growth in Discovered Oil and Gas Fields in Major Countries

The future reserve growth in discovered oil and gas fields in Turkmenistan rank first in Central Asia, accounting for 61.7% of the total reserve growth in Central Asia and mainly consists of gas. The future reserve growth from discovered oil and gas fields in Kazakhstan accounts for 22.0% of the total reserve growth in Central Asia. In discovered oil and gas fields in this country, the future reserve growth of oil are more than those of gas, which account for 67.7% of the total reserve growth of oil and gas. Other countries rank as follows in terms of future reserve growth from discovered oil and gas fields (in descending order): Azerbaijan, Uzbekistan, Tajikistan, Kyrgyzstan, Georgia (Figs. 9.6 and 9.7).

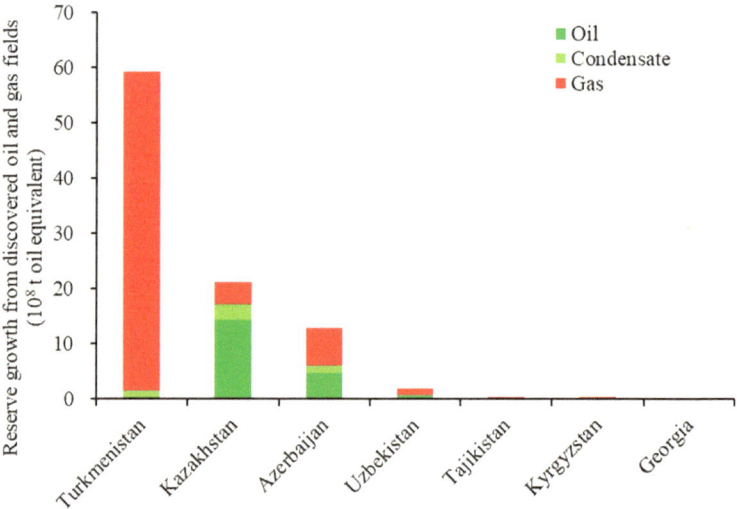

Fig. 9.6 Histogram of future reserve growth in discovered oil and gas fields in major Central Asian countries

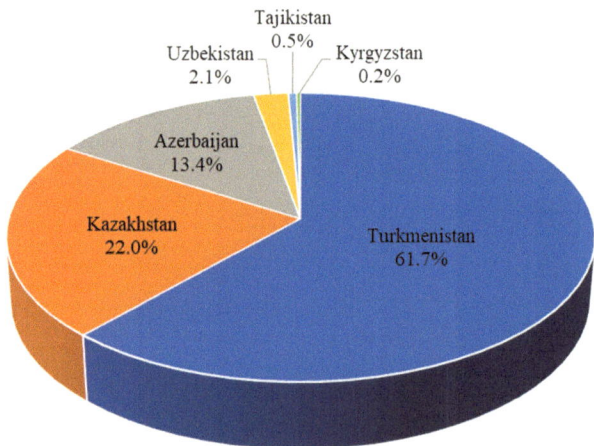

Fig. 9.7 Pie chart of future reserve growth in discovered oil and gas fields in major Central Asian countries

9.1.2.2   Distribution of Reserve Growth in Discovered Oil and Gas Fields in Major Basins

96.0% of the future reserve growth in discovered oil and gas fields in Central Asia are concentrated in the top four basins, namely, the Amu Darya Basin, the Precaspian Basin, the South Caspian Basin, and the North Ustyurt Basin (Figs. 9.8 and 9.9).

The future reserve growth in the Amu Darya Basin is 59.2×10^8 tons of oil equivalent, of which oil accounts for 0.8%, condensate accounts for 1.8%, and gas accounts for 97.4%. The future reserve growth in the Precaspian Basin is $17.3 \times$

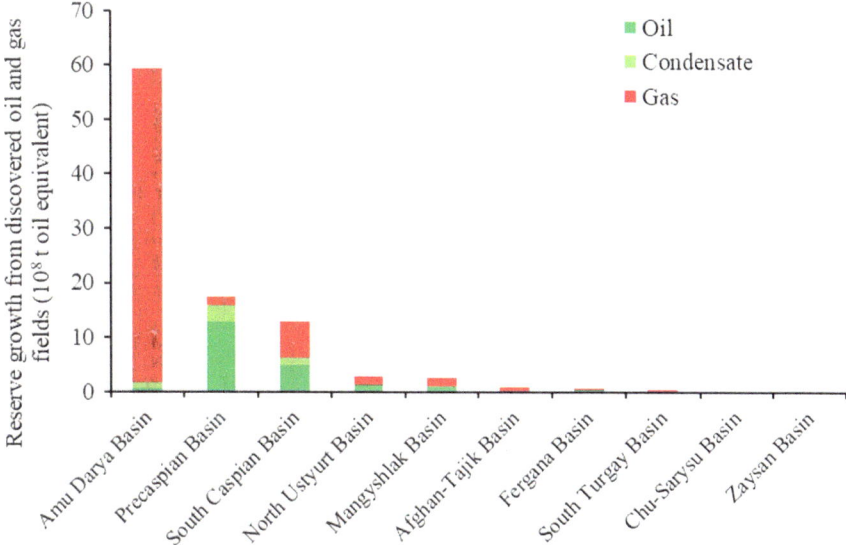

Fig. 9.8 Histogram of future reserve growth in discovered oil and gas fields in major basins in Central Asian

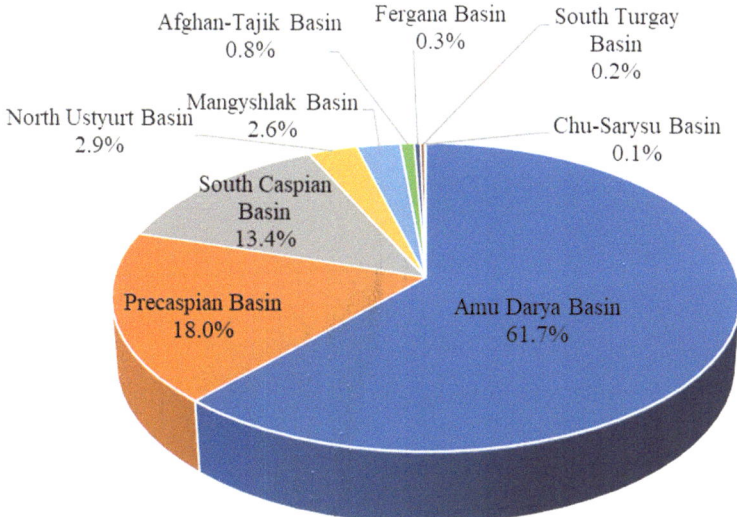

Fig. 9.9 Pie chart of future reserve growth in discovered oil and gas fields in major basins in Central Asian

10^8 tons of oil equivalent, of which oil accounts for 74.5%, condensate accounts for 16.4%, and gas accounts for 9.1%. The future reserve growth in the South Caspian Basin is 12.8×10^8 tons of oil equivalent, of which oil accounts for 37.1%, condensate accounts for 11.1%, and gas accounts for 51.8%.

9.1.2.3 Distribution of Reserve Growth in Discovered Onshore and Offshore Oil and Gas Fields

In Central Asia, the future reserve growth in discovered onshore oil and gas fields is much more than those in discovered offshore oil and gas fields. The future reserve growth in onshore and offshore fields accounts for 85.3% and 14.7%, respectively (Fig. 9.10). The future reserve growth in discovered onshore oil and gas fields amount to 81.9×10^8 tons of oil equivalent, of which oil accounts for 46.4%, condensate accounts for 4.6%, and gas accounts for 49.0%. The future reserve growth in discovered offshore oil and gas fields are 14.1×10^8 tons of oil equivalent, of which oil accounts for 19.4%, condensate accounts for 4.7%, and gas accounts for 75.9%.

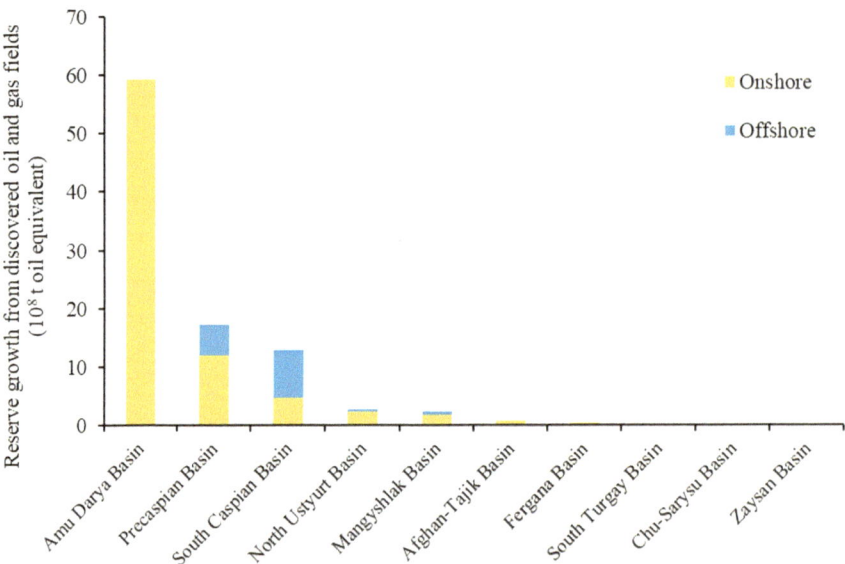

Fig. 9.10 Histogram of future reserve growth in discovered onshore and offshore oil and gas fields in major basins in Central Asia

9.1.3 Characteristics of Distributions of Undiscovered Oil and Gas Resources

The undiscovered oil and gas resources in Central Asia amount to 283.2×10^8 tons of oil equivalent, including 55.0×10^8 tons of oil, accounting for 19.4%, 12.7×10^8 tons of condensate, accounting for 4.5%, and 25.2×10^{12} m^3 of gas, accounting for 72.8%.

9.1.3.1 Distributions of Undiscovered Oil and Gas Resources in Major Countries

Over 92.7% of the undiscovered oil and gas resources in Central Asia are concentrated in four countries, namely, Turkmenistan, Kazakhstan, Azerbaijan, and Uzbekistan. Turkmenistan ranks first in Central Asia in terms of its undiscovered oil and gas resources, which account for about 48.3% of the total undiscovered oil and gas resources in Central Asia and mainly consist of gas. The country ranking second in terms of undiscovered oil and gas resources is Kazakhstan, in which the undiscovered gas resources are slightly more than the undiscovered oil resources. The country ranking third is Azerbaijan, in which the undiscovered oil resources are slightly more than the undiscovered gas resource. Other countries rank as follows (in descending order): Uzbekistan, Kyrgyzstan, Georgia, Tajikistan (Figs. 9.11 and 9.12).

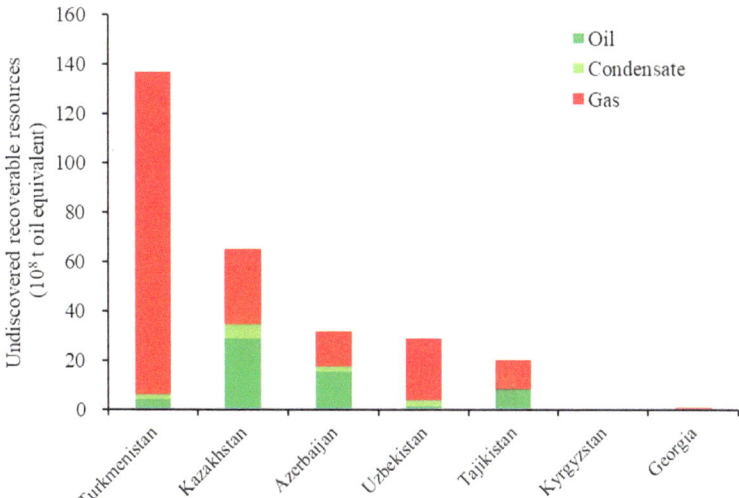

Fig. 9.11 Histogram of undiscovered recoverable oil and gas resources distributed in major Central Asian countries

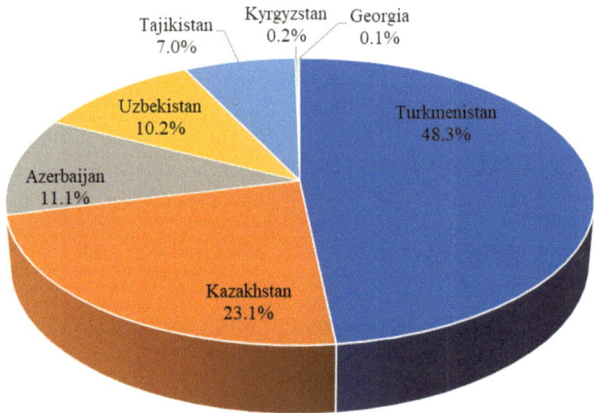

Fig. 9.12   Pie chart of undiscovered recoverable oil and gas resources distributed in major Central Asian countries

9.1.3.2   Distributions of Undiscovered Oil and Gas Resources in Major Basins

97.9% of the undiscovered oil and gas resources in Central Asia are concentrated in the top ten basins (Fig. 9.13). The Amu Darya Basin, Precaspian Basin and South Caspian Basin are the top three basins ranked in terms of undiscovered oil and gas resources. The sum of undiscovered oil and gas resources in these three basins accounts for 81.2% of the total undiscovered oil and gas resources in Central Asia (Fig. 9.14).

The undiscovered recoverable oil and gas resources in the Amu Darya Basin amount to 151.9×10^8 tons of oil equivalent, of which oil accounts for 0.4%, condensate accounts for 2.9%, and gas accounts for 96.8%. The undiscovered recoverable oil and gas resources in the Precaspian Basin are 48.4×10^8 tons of oil equivalent, of which oil accounts for 47.5%, condensate accounts for 11.3%, and gas accounts for 41.2%. The undiscovered recoverable oil and gas resources in the South Caspian Basin are 29.5×10^8 tons of oil equivalent, of which oil accounts for 48.0%, condensate accounts for 5.9%, and gas accounts for 46.1%.

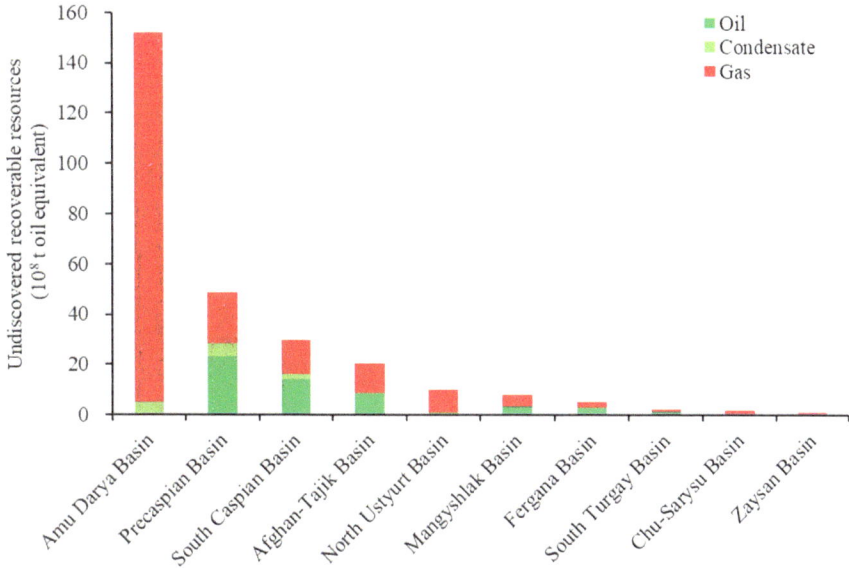

Fig. 9.13 Histogram of undiscovered recoverable oil and gas resources distributed in major basins in Central Asia

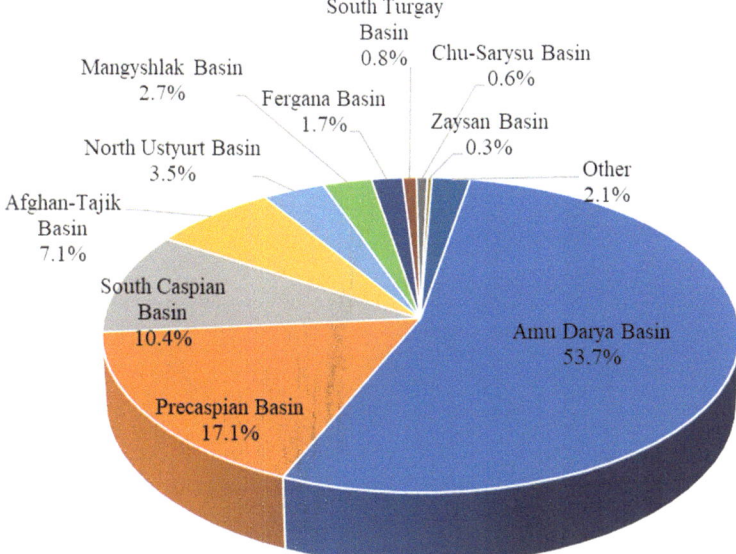

Fig. 9.14 Pie chart of undiscovered recoverable oil and gas resources distributed in major basins in Central Asia

9.1.3.3 Distributions of Undiscovered Oil and Gas Resources in Onshore and Offshore Areas

In Central Asia, the undiscovered recoverable oil and gas resources in onshore areas are much more than those in offshore areas. The undiscovered recoverable oil and gas resources in onshore and offshore areas account for 84.9% and 15.1%, respectively (Fig. 9.15). The undiscovered onshore gas resources are more than the undiscovered onshore oil resources, while the undiscovered offshore gas resources are slightly less than the undiscovered offshore oil resources.

The undiscovered recoverable oil and gas resources in onshore areas of Central Asia are 240.0×10^8 tons of oil equivalent, of which oil accounts for 13.8%, condensate accounts for 4.3%, and gas accounts for 81.9%. The undiscovered recoverable oil and gas resources in offshore areas are 43.3×10^8 tons of oil equivalent, of which oil accounts for 47.4%, condensate accounts for 7.6%, and gas accounts for 45.0%.

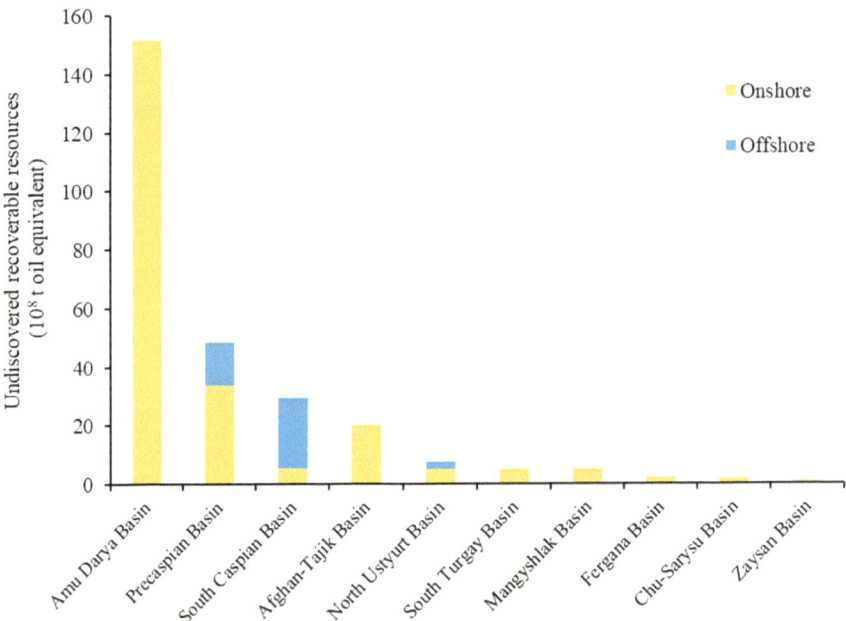

Fig. 9.15 Histogram of undiscovered recoverable oil and gas resources in onshore and offshore areas of major basins in Central Asia

9.2 Unconventional Oil and Gas Resources

The unconventional oil and gas resources in Central Asia include heavy oil, oil sands, shale oil, and shale gas. The recoverable unconventional resources in Central Asia are 207.6×10^8 tons of oil equivalent, accounting for 2.9% of the world's total recoverable unconventional resources.

The technically recoverable unconventional oil resources in Central Asia are 120.6×10^8 tons, accounting for 2.8% of the world's total recoverable unconventional oil resources. Among the total recoverable unconventional oil resources in Central Asia, recoverable oil sands are 59.4×10^8 tons, accounting for 49.3%, recoverable heavy oil resources are 44.6×10^8 tons, accounting for 37.0%, and recoverable shale oil resources are 16.6×10^8 tons, accounting for 13.7% (Figs. 9.16 and 9.17). The technically recoverable unconventional gas resources in Central Asia, all of which are shale gas, amount to 10.2×10^{12} m^3, accounting for 2.9% of the total recoverable unconventional gas resources worldwide.

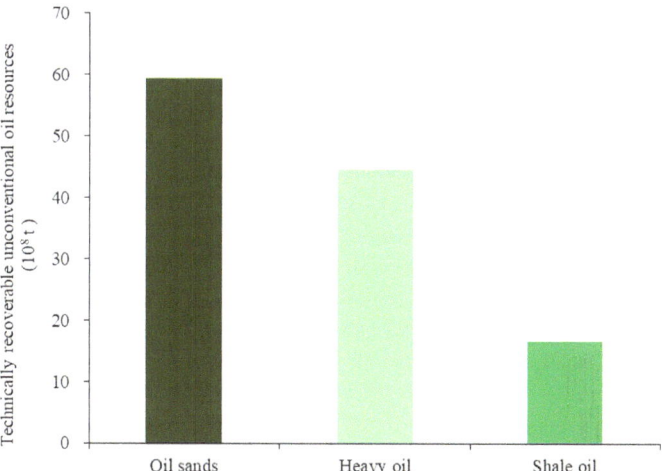

Fig. 9.16 Histogram of technically recoverable unconventional oil resources in Central Asia

Fig. 9.17 Pie chart of
technically recoverable
unconventional oil resources
in Central Asia

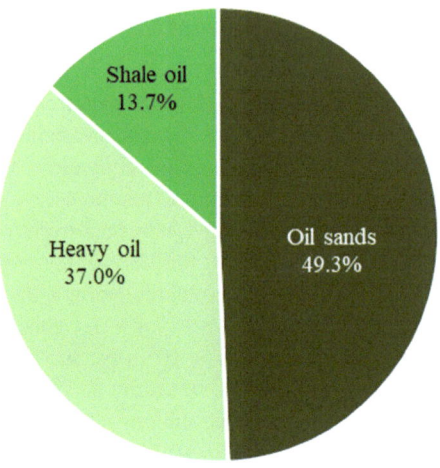

9.2.1 *Distribution of Recoverable Unconventional Oil and Gas Resources in Major Countries (Regions)*

9.2.1.1 Distribution of Recoverable Unconventional Oil Resources in Major Countries (Regions)

The unconventional oil resources in Central Asia are mainly distributed in three countries, namely, Kazakhstan, Georgia, and Turkmenistan. Kazakhstan ranks first in terms of its recoverable unconventional oil resources, which amount to 84.3 $\times$ 10^8 tons and account for 69.9% of the total recoverable unconventional oil resources in Central Asia and 2.1% of the world's total recoverable unconventional oil resources. Kazakhstan's unconventional oil resources, as ranked in descending order, include heavy oil, oil sands, and shale oil. The recoverable heavy oil, oil sands, and shale oil in Kazakhstan are 34.3 $\times$ 10^8 tons, 33.4 $\times$ 10^8 tons, and 16.6 $\times$ 10^8 tons, respectively. The recoverable unconventional oil resources in Georgia are 31.4 $\times$ 10^8 tons, accounting for 26.0% of the total recoverable unconventional oil resources in Central Asia and 0.8% of the world's total recoverable unconventional oil resources. The unconventional oil resources in Georgia are mainly heavy oil and oil sands, which amount to 6.8 $\times$ 10^8 tons and 24.6 $\times$ 10^8 tons, respectively. The recoverable unconventional oil resources in Turkmenistan are 4.9 $\times$ 10^8 tons, accounting for 4.0% of the total recoverable unconventional oil resources in Central Asia and 0.1% of the world's total recoverable unconventional oil resources (Figs. 9.18 and 9.19).

9.2.1.2 Distribution of Recoverable Unconventional Gas Resources in Major Countries (Regions)

The unconventional gas resources in Central Asia are mainly concentrated in Kazakhstan and Turkmenistan. The recoverable unconventional gas resources in these

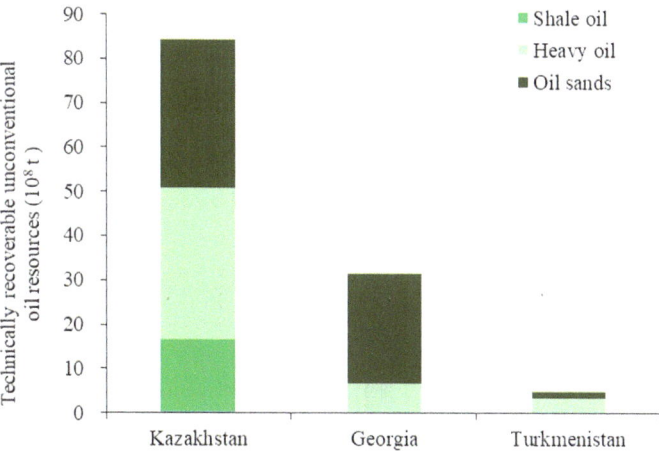

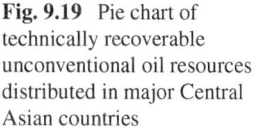

Fig. 9.18 Histogram of technically recoverable unconventional oil resources distributed in major Central Asian countries

Fig. 9.19 Pie chart of technically recoverable unconventional oil resources distributed in major Central Asian countries

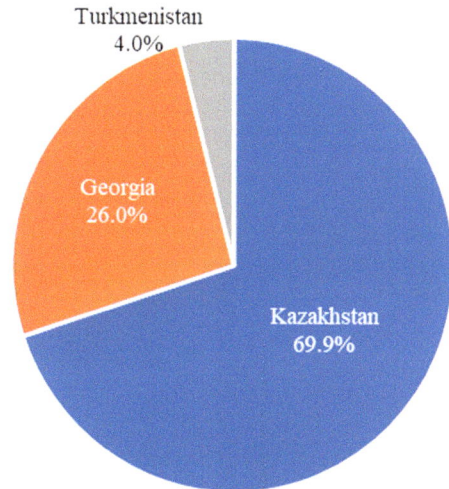

two countries are mainly shale gas resources. The recoverable shale gas resources in Kazakhstan are 0.8×10^{12} m^3, accounting for 29.0% of the total recoverable shale gas resources in Central Asia. The recoverable shale gas resources in Turkmenistan are 2.0×10^{12} m^3, accounting for 71.0% of the total recoverable shale gas resources in Central Asia.

9.2.2   Distribution of Recoverable Unconventional Oil and Gas Resources in Major Basins

The recoverable unconventional oil and gas resources in Central Asia are mainly distributed in the Precaspian Basin, North Ustyurt Basin, South Caspian Basin, Amu Darya Basin, Mangyshlak Basin, and South Turgay Basin.

9.2.2.1   Distribution of Recoverable Unconventional Oil Resources in Major Basins

The recoverable unconventional oil resources in Central Asia are mainly distributed in six basins, among which the Precaspian Basin, North Ustyurt Basin and South Caspian Basin rank among the top three. The recoverable unconventional oil resources in the Precaspian Basin are 43.3×10^8 tons, accounting for 35.9% of the total recoverable unconventional oil resources in Central Asia and including 9.9 $\times 10^8$ tons of shale oil and 33.4×10^8 tons of oil sands. The recoverable unconventional oil resources in the North Ustyurt Basin are 36.0×10^8 tons, accounting for 29.9% of the total recoverable unconventional oil resources in Central Asia and including 1.7×10^8 tons of shale oil and 34.3×10^8 tons of heavy oil. The recoverable unconventional oil resources in the South Caspian Basin are 31.4×10^8 tons, accounting for 26.0% of the total recoverable unconventional oil resources in Central Asia and including 6.8×10^8 tons of heavy oil and 24.6×10^8 tons of oil sands. The other basins rank as follows in terms of recoverable unconventional oil resources (in descending order): Amu Darya Basin, South Turgay Basin, Mangyshlak Basin (Figs. 9.20 and 9.21).

9.2.2.2   Distribution of Recoverable Unconventional Gas Resources in Major Basins

The unconventional gas resources in Central Asia, which are mainly shale gas, are concentrated in four basins. The Amu Darya Basin ranks first in terms of its recoverable unconventional gas resources, which amount to 7.6×10^{12} m^3 and account for 75.2% of the total recoverable unconventional gas resources in Central Asia. The South Turgay Basin ranks second in terms of its recoverable unconventional gas resources, which amount to 1.38×10^{12} m^3 and account for 13.6% of the total recoverable unconventional gas resources in Central Asia. The Mangyshlak Basin ranks third in terms of its recoverable shale gas resources, which amount to 0.59×10^{12} m^3 and account for 5.8% of the total recoverable shale gas resources in Central Asia. The recoverable shale gas resources in the South Turgay Basin are 0.58×10^{12} m^3 and account for 5.7% of the total recoverable shale gas resources in Central Asia (Figs. 9.22 and 9.23).

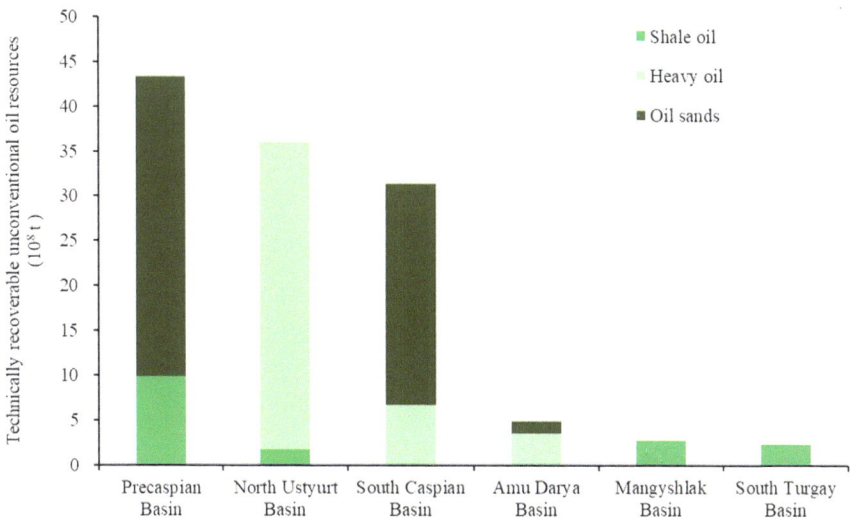

Fig. 9.20 Histogram of technically recoverable unconventional oil resources in major basins in Central Asia

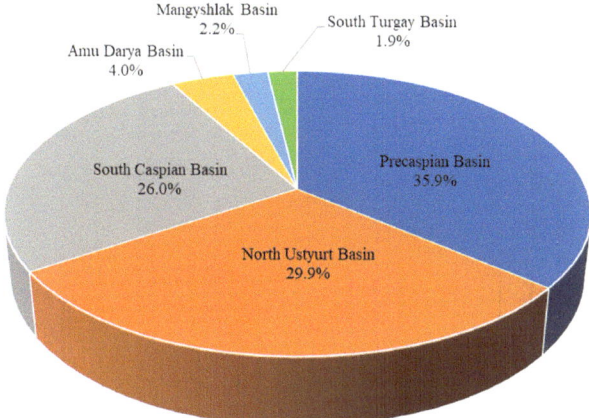

Fig. 9.21 Pie chart of technically recoverable unconventional oil resources in major basins in Central Asia

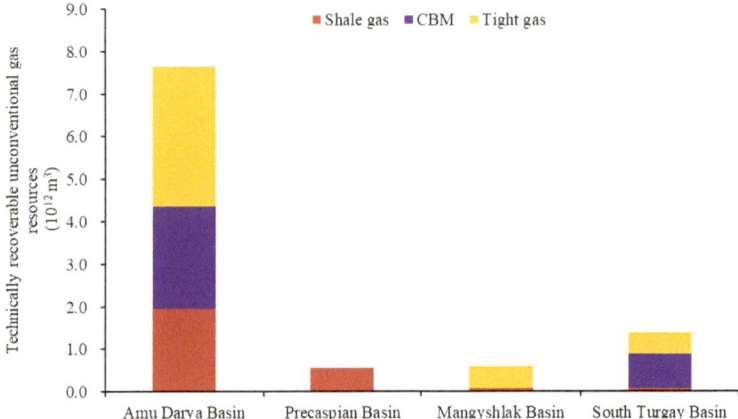

Fig. 9.22 Histogram of technically recoverable unconventional gas resources in major basins in Central Asia

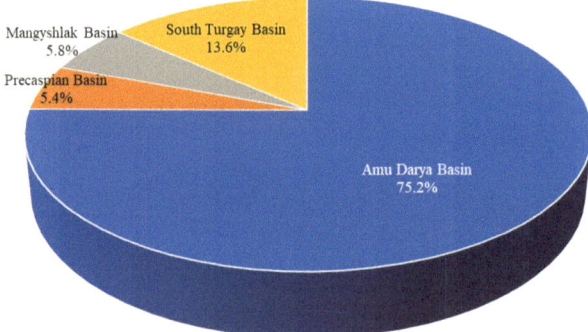

Fig. 9.23 Pie chart of technically recoverable unconventional gas resources in major basins in Central Asia

Chapter 10
Distribution of Oil and Gas Resources in Russia

Russia is located in the northern part of Eurasian continent. Twenty-six oil and gas-bearing basins have developed in this region, covering a land area of 1709.82×10^4 km^2. Onshore basins include cratonic basins, rift basins, and foreland basins. The basins in Arctic seas are mainly passive margin basins, and the basins in the Western Pacific are mainly back-arc basins. 14.8% of the world's oil and gas resources are concentrated in Russia. The total recoverable oil and gas resources in Russia amount to 2773.2×10^8 tons of oil equivalent.

10.1 Conventional Oil and Gas Resources

The recoverable conventional oil and gas resources in Russia amount to 1755.9×10^8 tons of oil equivalent, accounting for 15.4% of the world's total recoverable conventional resources. The recoverable reserves of oil and gas in Russia amount to 1029.5×10^8 tons of oil equivalent, accounting for 15.8% of the world's total recoverable reserves. The cumulative oil and gas production in Russia is 468.8×10^8 tons of oil equivalent, accounting for 18.0% of the world's total cumulative oil and gas production. The remaining recoverable reserves of oil and gas in Russia are 560.6×10^8 tons of oil equivalent, accounting for 14.3% of the world's total remaining recoverable reserves. The predicted reserve growth from discovered oil and gas fields in Russia are 132.5×10^8 tons of oil equivalent, accounting for 11.8% of the world's total reserve growth in the future. The undiscovered recoverable oil and gas resources in Russia are 594.0×10^8 tons of oil equivalent, accounting for 15.6% of the world's total undiscovered recoverable resources.

© The Author(s) 2024
L. Dou et al., *Global Oil and Gas Resources: Potential and Distribution*,
https://doi.org/10.1007/978-981-97-4756-6_10

10.1.1   *Distribution of Remaining Recoverable Reserves*

The remaining recoverable reserves of oil and gas in Russia are 560.6×10^8 tons of oil equivalent, including 194.6×10^8 tons of oil, accounting for 34.7%, 24.2×10^8 tons of condensate, accounting for 4.3%, and 40×10^{12} m^3 of gas, accounting for 61.0%.

10.1.1.1   Distribution of Remaining Recoverable Reserves in Various Basins

Almost all the remaining recoverable reserves of oil and gas in Russia are distributed in the top ten basins (Fig. 10.1). The West Siberian Basin, Volga-Ural Basin, and East Siberian Basin rank among the top three in terms of remaining recoverable reserves. The sum of remaining recoverable reserves in these three basins accounts for 81.5% of the total remaining recoverable reserves in Russia (Fig. 10.2).

The remaining recoverable reserves in the West Siberian Basin amount to 364.0 $\times 10^8$ tons of oil equivalent, of which oil accounts for 32.7%, condensate accounts for 3.7%, and gas accounts for 63.6%. The remaining recoverable reserves in the Volga-Ural Basin are 47.2×10^8 tons of oil equivalent, of which oil accounts for 86.5%, condensate accounts for 0.7%, and gas accounts for 12.8%. The remaining recoverable reserves in the East Siberian Basin are 45.6×10^8 tons of oil equivalent, of which oil accounts for 24.6%, condensate accounts for 3.6%, and gas accounts for 71.8%.

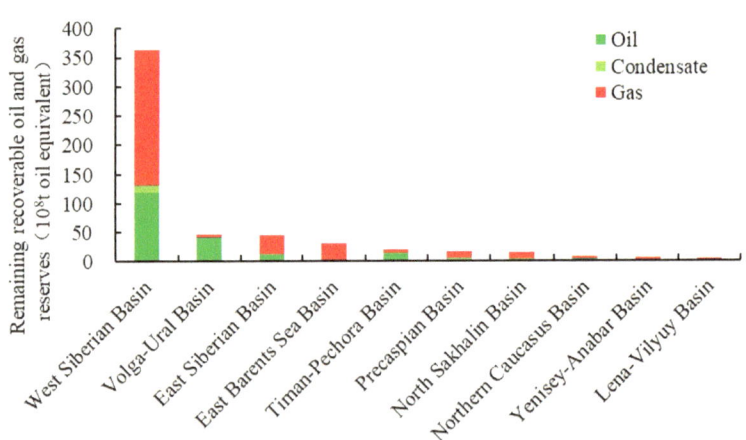

Fig. 10.1 Histogram of the remaining recoverable reserves in top ten basins in Russia

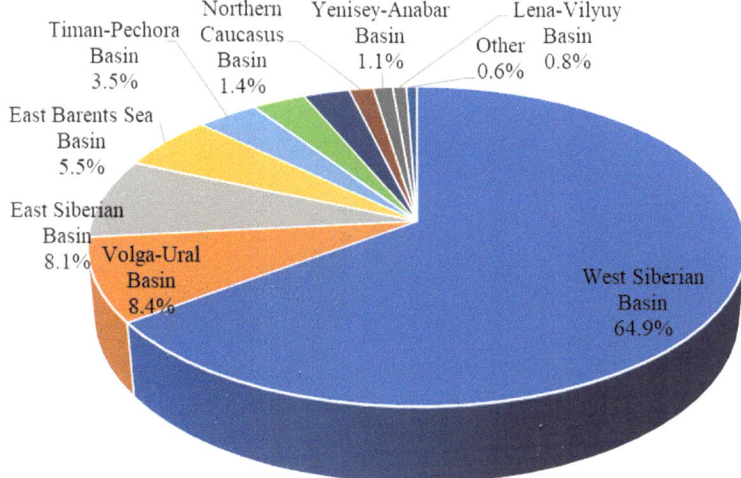

Fig. 10.2 Pie chart of the remaining recoverable reserves in major basins in Russia

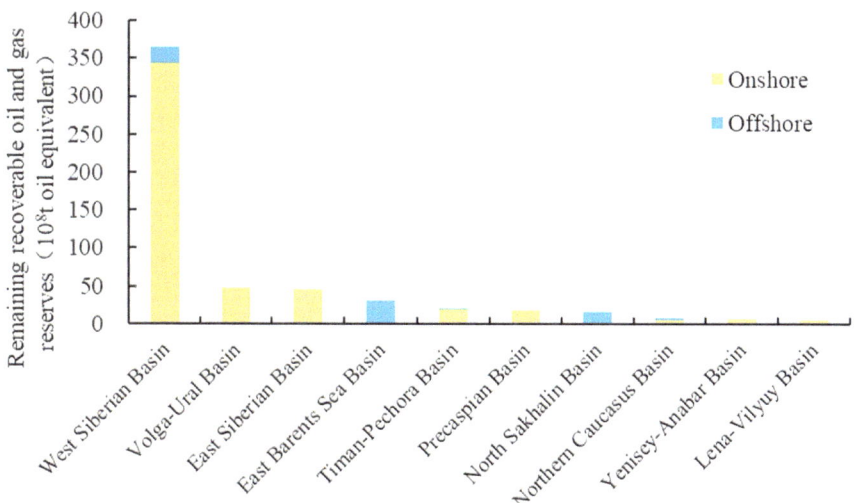

Fig. 10.3 Histogram of remaining recoverable reserves distributed in major basins in Russia

10.1.1.2 Distribution of Remaining Recoverable Reserves in Onshore and Offshore Areas

In Russia, the remaining recoverable oil and gas reserves in onshore areas are much more than those in offshore areas. The remaining recoverable reserves in onshore and offshore areas of Russia account for 86.7% and 13.3%, respectively (Fig. 10.3).

The remaining recoverable reserves in onshore areas of Russia amount to 485.9 × 10^8 tons of oil equivalent, of which oil accounts for 39.0%, condensate accounts for 4.0%, and gas accounts for 57.0%. The remaining recoverable reserves in offshore areas of Russia are 74.7 × 10^8 tons of oil equivalent, of which oil accounts for 8.8%, condensate accounts for 4.4%, and gas accounts for 86.8%.

10.1.2   *Trend of Reserve Growth from Discovered Oil and Gas Fields*

The predicted reserve growth in discovered oil and gas fields across Russia are 132.5 × 10^8 tons of oil equivalent, including 63.8 × 10^8 tons of oil, accounting for 48.1%, 3.7 × 10^8 tons of condensate, accounting for 2.8%, and 7.6 × 10^{12} m^3 of gas, accounting for 49.1%.

10.1.2.1   Distribution of Reserve Growth in Discovered Oil and Gas Fields in Various Basins

99.9% of the future reserve growth in discovered oil and gas fields across Russia are concentrated in the top ten basins (Fig. 10.4). The West Siberian Basin, East Barents Sea Basin, and Volga-Ural Basin rank among the top three in terms of future reserve growth. The future reserve growth in discovered oil and gas fields in these three basins account for 88.1% of the total future reserve growth in Russia (Fig. 10.5).

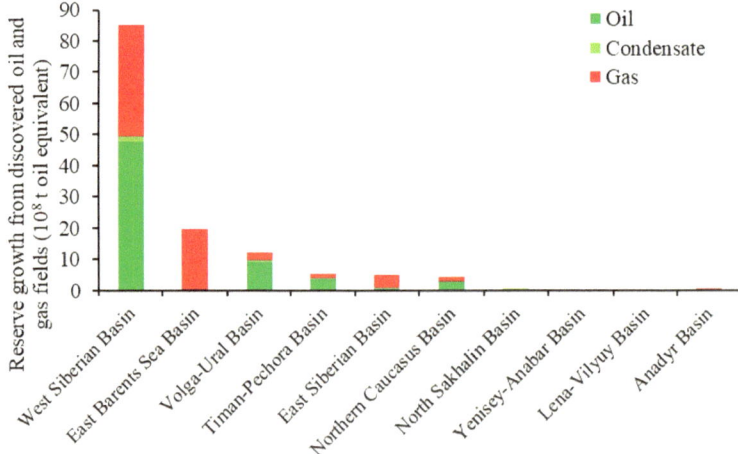

Fig. 10.4 Histogram of future reserve growth in discovered oil and gas fields in major basins in Russia

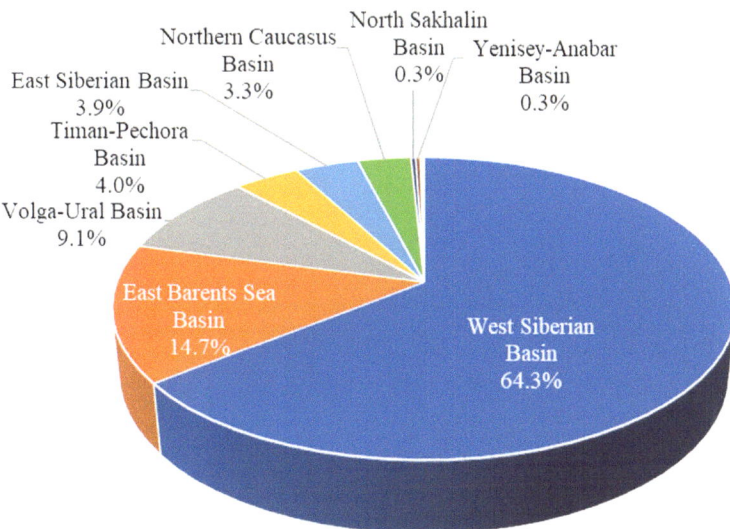

Fig. 10.5 Pie chart of future reserve growth in discovered oil and gas fields in major basins in Russia

The future reserve growth in discovered oil and gas fields in the West Siberian Basin amount to 85.2×10^8 tons of oil equivalent, of which oil accounts for 55.8%, condensate accounts for 2.0%, and gas accounts for 42.2%. The future reserve growth in the East Barents Sea Basin is 19.5×10^8 tons of oil equivalent, of which condensate accounts for 1.2% and gas accounts for 98.8%. The future reserve growth in the Volga-Ural Basin are 12.0×10^8 tons of oil equivalent, of which oil accounts for 76.2%, condensate accounts for 4.8%, and gas accounts for 19.0%.

10.1.2.2 Distribution of Reserve Growth in Discovered Onshore and Offshore Oil and Gas Fields

In Russia, the future reserve growth in discovered onshore oil and gas fields are much more than those in discovered offshore oil and gas fields. The future reserve growth in onshore and offshore areas of Russia account for 92.7% and 7.3%, respectively (Fig. 10.6). In both onshore and offshore areas, the additional gas reserves are more than the additional oil reserves.

The future reserve growth in discovered onshore oil and gas fields are 122.8×10^8 tons of oil equivalent, of which oil accounts for 45.8%, condensate accounts for 2.5%, and gas accounts for 51.7%. The future reserve growth in discovered offshore oil and gas fields are 9.7×10^8 tons of oil equivalent, of which oil accounts for 10.7%, condensate accounts for 4.4%, and gas accounts for 84.9%.

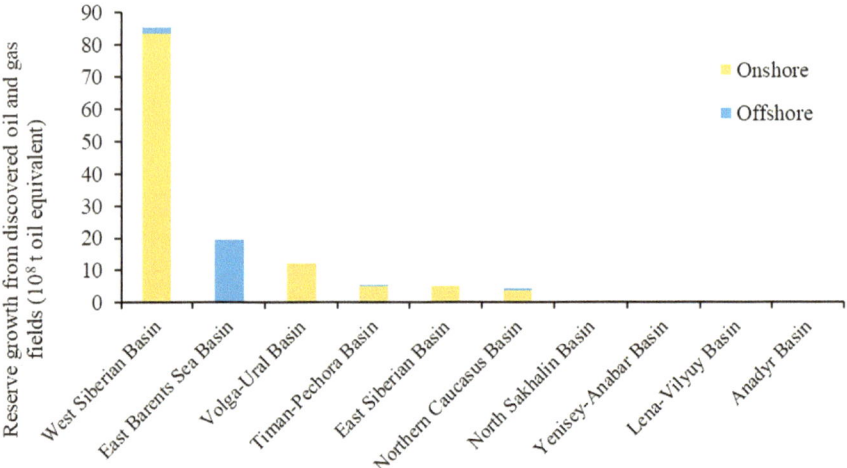

Fig. 10.6 Histogram of future reserve growth in onshore and offshore oil and gas fields in major basins in Russia

10.1.3   Characteristics of Distributions of Undiscovered Oil and Gas Resources

The undiscovered oil and gas resources in Russia amount to 594.0×10^8 tons of oil equivalent, including 148.8×10^8 tons of oil, accounting for 25.1%, 38.4×10^8 tons of condensate, accounting for 6.5%, and 47.6×10^{12} m^3 of gas, accounting for 68.4%.

10.1.3.1   Distribution of Undiscovered Oil and Gas Resources in Various Basins

96.8% of the undiscovered oil and gas resources in Russia are concentrated in the top ten basins (Fig. 10.7). The West Siberian Basin, East Siberian Basin, and East Barents Sea Basin rank among the top three in terms of undiscovered oil and gas resources. The undiscovered oil and gas resources in these three basins account for 78.8% of the total undiscovered oil and gas resources in Russia (Fig. 10.8).

The undiscovered oil and gas resources in the West Siberian Basin amount to 261.1 $\times 10^8$ tons of oil equivalent, of which oil accounts for 35.3%, condensate accounts for 7.2%, and gas accounts for 57.5%. The undiscovered oil and gas resources in the East Siberian Basin are 113.7×10^8 tons of oil equivalent, of which oil accounts for 10.8%, condensate accounts for 5.2%, and gas accounts for 84.0%. The undiscovered oil and gas resources in the East Barents Sea Basin are 93.2×10^8 tons of oil equivalent, of which oil accounts for 8.3%, condensate accounts for 4.6%, and gas accounts for 87.1%.

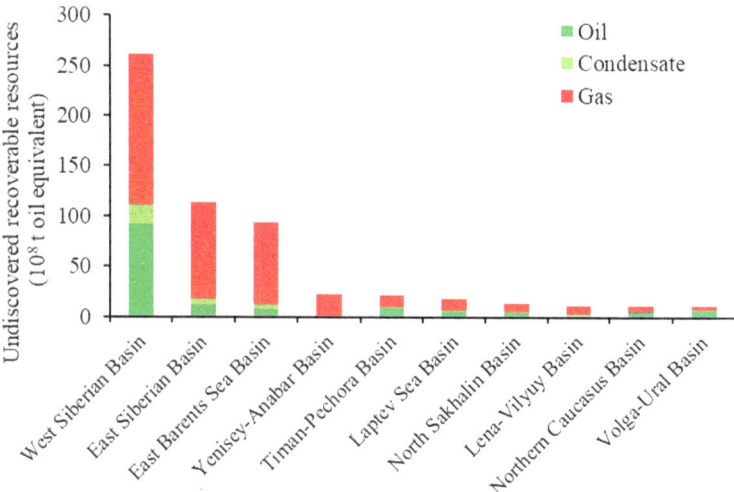

Fig. 10.7 Histogram of undiscovered recoverable oil and gas resources in major basins in Russia

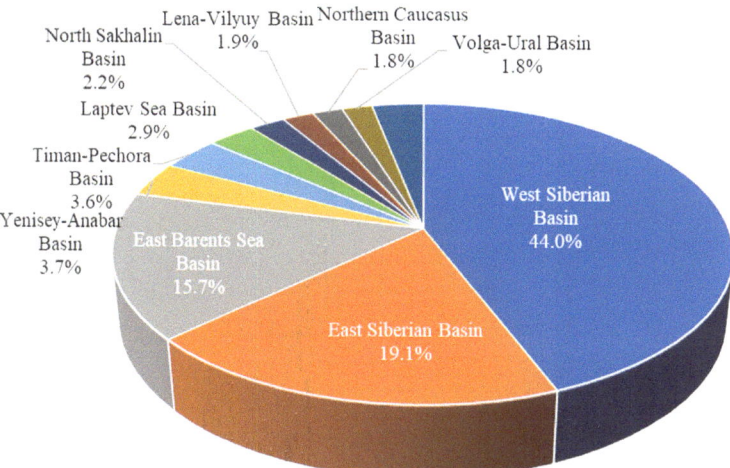

Fig. 10.8 Pie chart of undiscovered recoverable oil and gas resources in major basins in Russia

10.1.3.2 Distribution of Undiscovered Oil and Gas Resources in Onshore and Offshore Areas

In Russia, the undiscovered oil and gas resources in onshore areas are less than those in offshore areas. The undiscovered oil and gas resources in onshore and offshore areas of Russia account for 49.0% and 51.0%, respectively (Fig. 10.9). In both onshore and offshore areas, the undiscovered gas resources are much more than the undiscovered oil resources.

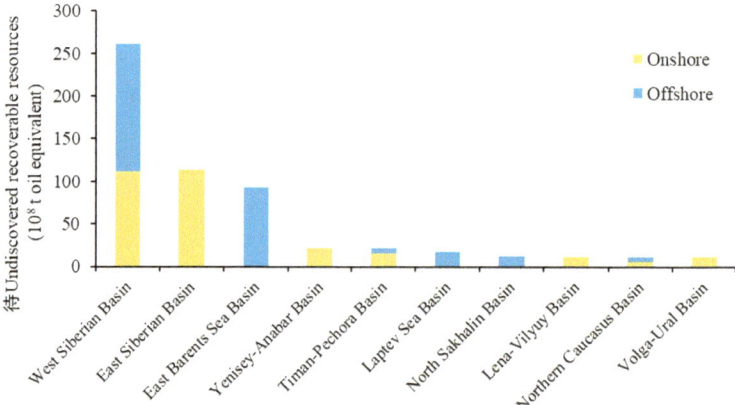

Fig. 10.9 Histogram of undiscovered recoverable oil and gas resources in onshore and offshore areas of major basins in Russia

The undiscovered oil and gas resources in onshore areas of Russia are 291.3×10^8 tons of oil equivalent, of which oil accounts for 23.3%, condensate accounts for 6.2%, and gas accounts for 70.5%. The undiscovered oil and gas resources in offshore areas of Russia are 303.7×10^8 tons of oil equivalent, of which oil accounts for 26.7%, condensate accounts for 6.7%, and gas accounts for 66.6%.

10.2   Unconventional Oil and Gas Resources

10.2.1   Distribution of Recoverable Unconventional Oil and Gas Resources

Russia has abundant unconventional oil and gas resources, including oil shale, heavy oil, oil sands, shale oil, shale gas, tight gas, and CBM. The recoverable unconventional oil and gas resources in Russia amount to 1017.3×10^8 tons of oil equivalent, accounting for 14% of the world's total recoverable unconventional resources.

10.2.1.1   Distribution of Recoverable Unconventional Oil Resources

The recoverable unconventional oil resources in Russia amount to 683.1×10^8 tons, accounting for 16.1% of the world's total recoverable unconventional oil resources. Among the recoverable unconventional oil resources in Russia, oil shale amount to 338.2×10^8 tons of oil equivalent, accounting for 49.5%, shale oil resources are 130.3×10^8 tons, accounting for 19.1%, oil sands amount to 125.7×10^8 tons,

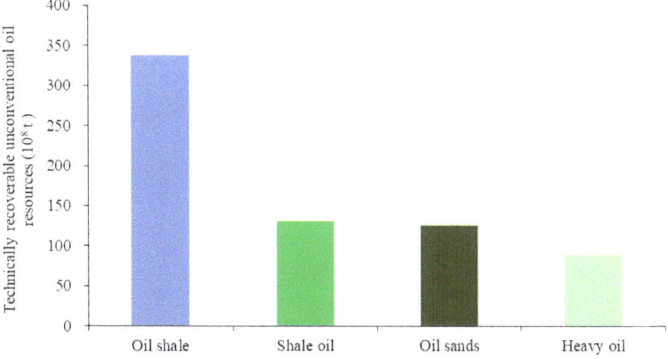

Fig. 10.10 Histogram of technically recoverable unconventional oil resources in Russia

Fig. 10.11 Pie chart of technically recoverable unconventional oil resources in Russia

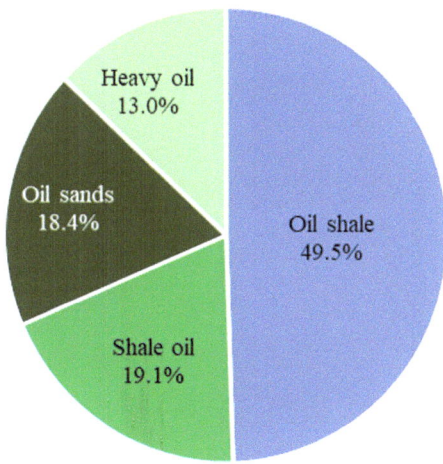

accounting for 18.4%, and heavy oil resources are 89.0×10^8 tons, accounting for 13.0% (Figs. 10.10 and 10.11).

10.2.1.2 Distribution of Recoverable Unconventional Gas Resources

The recoverable unconventional gas resources in Russia are 39.1×10^{12} m^3, which account for 11.1% of the world's total recoverable unconventional gas resources and include 19.7×10^{12} m^3 of shale gas, accounting for 50.4%, 13.1×10^{12} m^3 of CBM, accounting for 33.5%, and 6.3×10^{12} m^3 of tight gas, accounting for 16.1% (Figs. 10.12 and 10.13).

Fig. 10.12 Histogram of technically recoverable unconventional gas resources in Russia

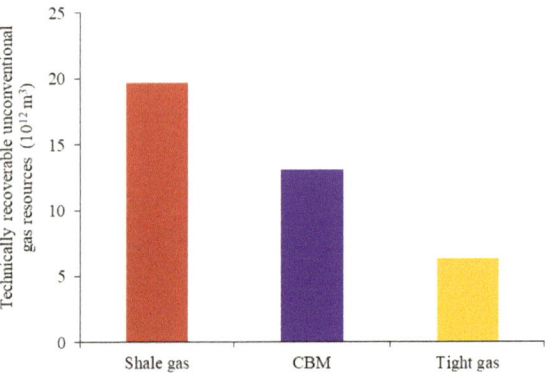

Fig. 10.13 Pie chart of technically recoverable unconventional gas resources in Russia

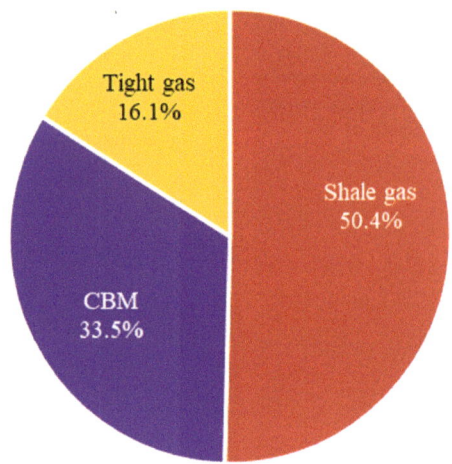

10.2.2 Distribution of Unconventional Oil and Gas Resources in Major Basins

The recoverable unconventional oil and gas resources in Russia are mainly distributed in six basins, including the West Siberian Basin, Timan-Pechora Basin, Volga-Ural Basin, and Northern Caucasus Basin on both sides of the Ural Mountains, the East Siberian Basin in the east, and the North Sakhalin Basin in the Far East. 83.5% of the recoverable unconventional oil resources in Russia are concentrated in the top three basins. The recoverable unconventional gas resources in Russia are mainly distributed in five basins, namely, the West Siberian Basin, the East Siberian Basin, the Volga-Ural Basin, the Timan-Pechora Basin, and the Kuznetsk Basin.

10.2.2.1 Distribution of Unconventional Oil Resources in Major Basins

The unconventional oil resources in Russia are mainly distributed in three basins, namely, the West Siberian Basin, the Volga-Ural Basin, and the East Siberian Basin. The recoverable unconventional oil resources in the West Siberian Basin amount to 229.9×10^8 tons, which account for 33.7% of the total recoverable unconventional oil resources in Russia and include 104.3×10^8 tons of oil shale, 93.2×10^8 tons of shale oil, and 32.4×10^8 tons of heavy oil. The recoverable unconventional oil resources in the Volga-Ural Basin amount to 190.1×10^8 tons, which account for 27.8% of the total recoverable unconventional oil resources in Russia and include 129.0×10^8 tons of oil shale, 42.2×10^8 tons of oil sands, and 18.8×10^8 tons of shale oil. The recoverable unconventional oil resources in the East Siberian Basin amount to 167.0×10^8 tons, which account for 24.5% of the total recoverable unconventional oil resources in Russia and include 104.8×10^8 tons of oil shale and 62.2×10^8 tons of oil sands. The other basins rank as follows in terms of recoverable unconventional oil resources (in descending order): Timan-Pechora Basin, Northern Caucasus Basin, North Sakhalin Basin, Anadyr Basin (Figs. 10.14 and 10.15).

In Russia, 71.6% of shale oil resources are concentrated in the West Siberian Basin, 68.3% of heavy oil resources are concentrated in the West Siberian Basin and Timan-Pechora Basin, 83.6% of oil sands are concentrated in the East Siberian Basin and Volga-Ural Basin, and oil shale are distributed in the Volga-Ural Basin, West Siberian Basin and East Siberian Basin.

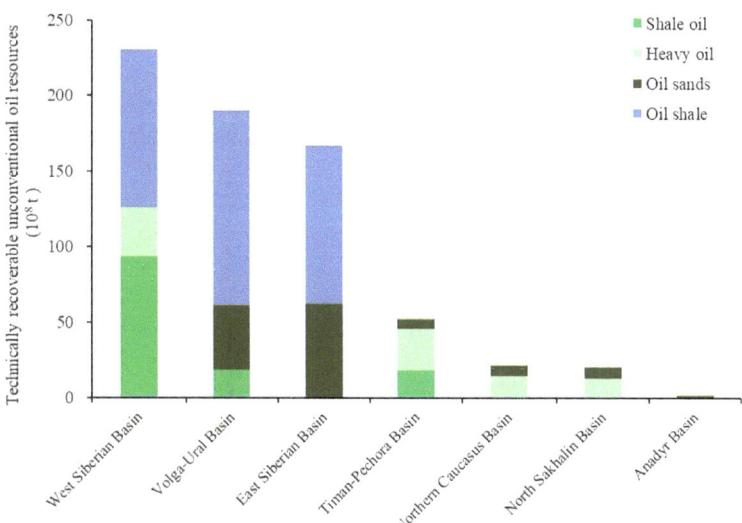

Fig. 10.14 Histogram of technically recoverable unconventional oil resources in major basins in Russia

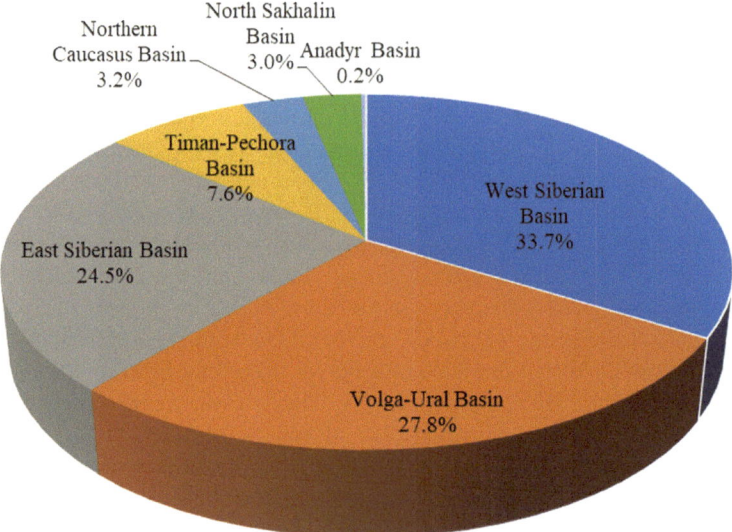

Fig. 10.15 Pie chart of technically recoverable unconventional oil resources in major basins in Russia

10.2.2.2 Distribution of Unconventional Gas Resources in Major Basins

The unconventional gas resources in Russia are mainly distributed in five basins, among which the East Siberian Basin, Kuznetsk Basin, and West Siberian Basin rank among the top three in terms of unconventional gas resources. The recoverable unconventional gas resources in the East Siberian Basin rank first in Russia and amount to 12.5×10^{12} m^3, accounting for 32.1% of the total recoverable unconventional gas resources in Russia and including 5.8×10^{12} m^3 of shale gas and 4.7×10^{12} m^3 of CBM, and 2.0×10^{12} m^3 of tight gas. The recoverable unconventional gas resources in the West Siberian Basin, which are mainly shale gas, rank second and amount to 10.2×10^{12} m^3, accounting for 26.2% of the total recoverable unconventional gas resources in Russia. The recoverable unconventional gas resources in the Kuznetsk Basin, which are mainly CBM, rank third and amount to 9.5×10^{12} m^3, accounting for 24% of the total recoverable unconventional gas resources in Russia. The other basins rank as follows in terms of recoverable unconventional gas resources (in descending order): Volga-Ural Basin, Timan-Pechora Basin (Figs. 10.16 and 10.17).

In Russia, 92.9% of shale gas resources are distributed in the West Siberian Basin, East Siberian Basin, and Volga-Ural Basin, CBM resources are concentrated in the Kuznetsk Basin and East Siberian Basin, and tight gas resources are concentrated in the Timan-Pechora Basin.

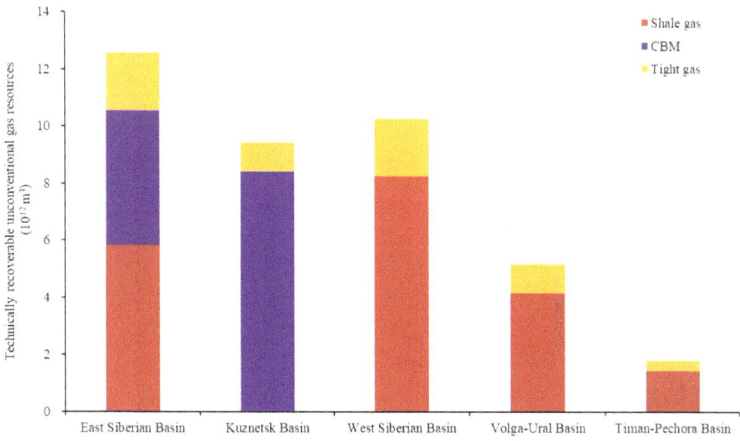

Fig. 10.16 Histogram of technically recoverable unconventional gas resources in major basins in Russia

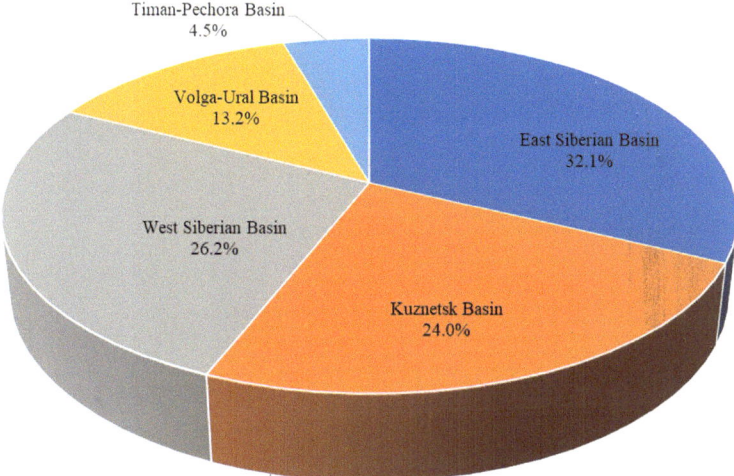

Fig. 10.17 Pie chart of technically recoverable unconventional gas resources in major basins in Russia

Chapter 11
Distribution of Oil and Gas Resources in the Asia–Pacific Region

The Asia–Pacific region refers to the Asian regions on the west coast of the Pacific Ocean, such as East Asia and Southeast Asia, Oceania, and the islands in the Pacific Ocean, including China and Japan in East Asia, Southeast Asia, as well as Australia and New Zealand. One hundred fourteen sedimentary basins have developed in the Asia–Pacific region, covering a total area of 3794×10^4 km^2. The onshore sedimentary basins have a total area of 1499×10^4 km^2, where the main types of basins are cratonic, foreland and rift basins. The offshore sedimentary basins have a total area of 2295×10^4 km^2, where the main types of basins are back-arc basins and passive margin basins. 12% of the world's oil and gas resources are concentrated in the Asia–Pacific region. The total recoverable oil and gas resources in the Asia–Pacific region amount to 2250.1×10^8 tons of oil equivalent.

11.1 Conventional Oil and Gas Resources

The recoverable conventional oil and gas resources in the Asia–Pacific region amount to 1377.6×10^8 tons of oil equivalent, accounting for 12% of the world's total recoverable conventional resources. The recoverable reserves of oil and gas in the Asia–Pacific region amount to 592.9×10^8 tons of oil equivalent, accounting for 9.1% of the world's total recoverable reserves. The cumulative oil and gas production in this region is 280.1×10^8 tons of oil equivalent, accounting for 10.7% of the world's total cumulative oil and gas production. The remaining recoverable reserves of oil and gas in this region are 312.8×10^8 tons of oil equivalent, accounting for 8.0% of the world's total remaining recoverable reserves. The predicted reserve growth from discovered oil and gas fields in this region are 108.4×10^8 tons of oil equivalent, accounting for 9.7% of the total estimated reserve growth from discovered oil and gas fields worldwide. The undiscovered recoverable oil and gas resources the Asia–Pacific region are 676.3×10^8 tons of oil equivalent, accounting for 17.8% of the world's total undiscovered recoverable resources.

11.1.1 Distribution of Remaining Recoverable Reserves

The remaining recoverable reserves of oil and gas in the Asia–Pacific region are 312.8×10^8 tons of oil equivalent, of which oil accounts for 23.4%, condensate accounts for 4.9%, and gas accounts for 71.7%.

11.1.1.1 Distribution of Remaining Recoverable Reserves in Major Countries

In terms of remaining recoverable reserves. China ranks first, and Indonesia and Australia rank second and third (Figs. 11.1 and 11.2).

The remaining recoverable reserves of oil and gas in China amount to 105.1×10^8 tons of oil equivalent, of which oil accounts for 38.5%, condensate accounts for 2.4%, and gas accounts for 59.1%. The remaining recoverable reserves in Indonesia are 53.6×10^8 tons of oil equivalent, of which oil accounts for 19.5%, condensate accounts for 4.8%, and gas accounts for 75.7%. The remaining recoverable reserves in Australia are 49.6×10^8 tons of oil equivalent, of which oil accounts for 3.8%, condensate accounts for 8.0%, and gas accounts for 88.2%.

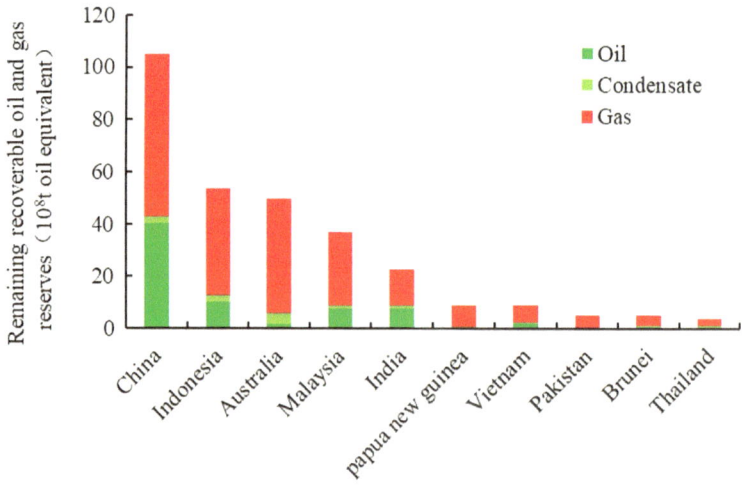

Fig. 11.1 Histogram of remaining recoverable reserves in major countries of the Asia–Pacific region

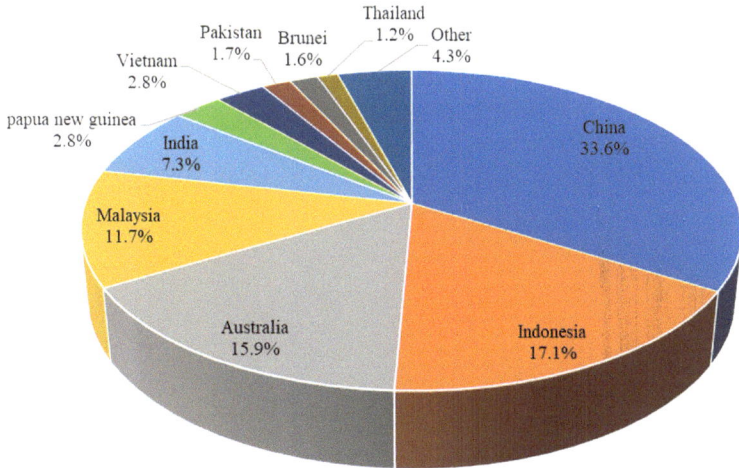

Fig. 11.2 Pie chart of remaining recoverable reserves in major countries of the Asia–Pacific region

11.1.1.2 Distribution of Remaining Recoverable Reserves in Major Basins

The remaining recoverable reserves of oil and gas in the Asia–Pacific region are mainly distributed in 84 basins, including the Zengmu Basin, North Carnarvon Basin, Ordos Basin, Tarim Basin, and Sichuan Basin, and so on. The sum of remaining recoverable reserves in the top 10 basins accounts for 58.1% of the total remaining recoverable reserves in the Asia–Pacific region (Figs. 11.3 and 11.4).

The remaining recoverable reserves of oil and gas in Zengmu Basin are 27.1×10^8 tons of oil equivalent, of which oil accounts for 8.5%, condensate accounts for

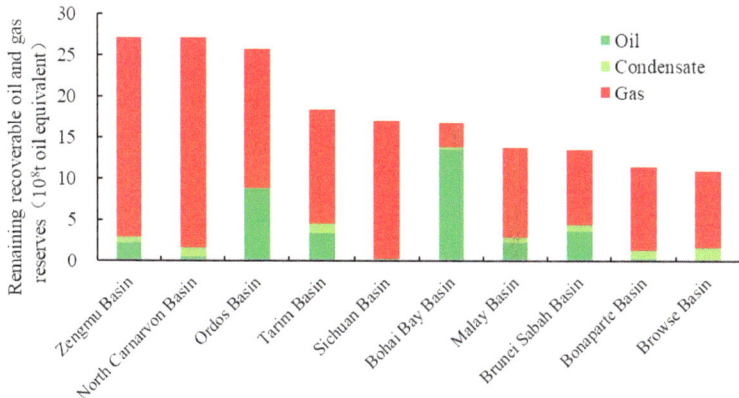

Fig. 11.3 Histogram of remaining recoverable reserves distributed in major basins of the Asia–Pacific region

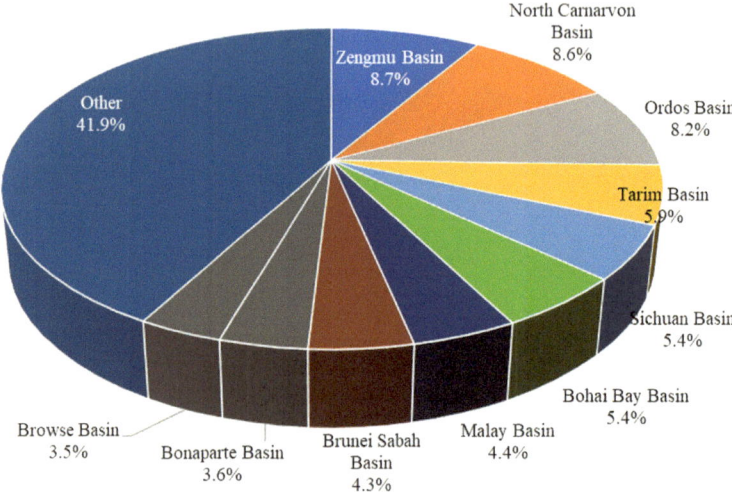

Fig. 11.4 Pie chart of remaining recoverable reserves distributed in major basins of the Asia–Pacific region

2.2%, and gas accounts for 89.3%. The remaining recoverable reserves in the North Carnarvon Basin rank second and amount to 27.0×10^8 tons of oil equivalent, of which oil accounts for 2.0%, condensate accounts for 4.0%, and gas accounts for 94%. The remaining recoverable reserves in the Ordos Basin are 25.7×10^8 tons of oil equivalent, of which oil accounts for 34%, condensate accounts for 0.6%, and gas accounts for 65.4%.

11.1.1.3   Distribution of Remaining Recoverable Reserves in Onshore and Offshore Areas

The remaining recoverable reserves of oil and gas in the offshore areas of the Asia–Pacific region account for 62% of the total remaining recoverable reserves in this region. In both onshore and offshore areas of this region, the remaining recoverable gas reserves are significantly more than the remaining recoverable Oil reserves (Fig. 11.5). The remaining recoverable reserves of offshore oil and gas in this region are 193.9×10^8 tons of oil equivalent, of which oil accounts for 19.6%, condensate accounts for 4.5%, and gas accounts for 75.9%. In contrast, the remaining recoverable reserves of onshore oil and gas are 118.9×10^8 tons of oil equivalent, of which oil accounts for 10.8%, condensate accounts for 6.5%, and gas accounts for 82.7%.

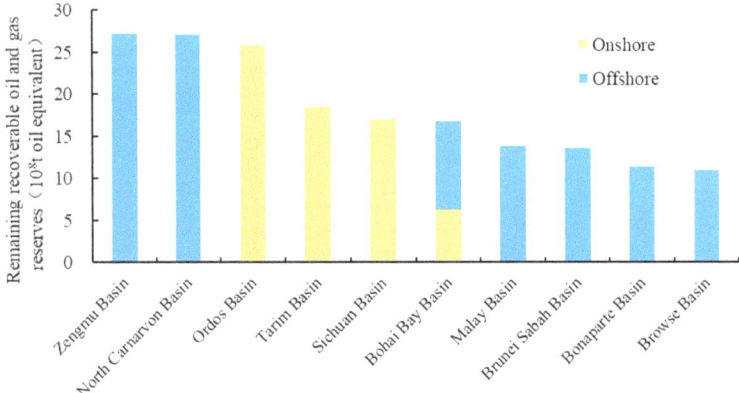

Fig. 11.5 Histogram of remaining recoverable reserves in onshore and offshore areas of major basins in the Asia–Pacific region

11.1.2 Trend of Reserve Growth from Discovered Oil and Gas Fields

The future reserve growth from discovered oil and gas fields in the Asia–Pacific region is 108.4×10^8 tons of oil equivalent, of which oil accounts for 22.7%, condensate accounts for 4.5%, and gas accounts for 72.8%.

11.1.2.1 Distribution of Reserve Growth in Discovered Oil and Gas Fields in Major Countries (Regions)

24.4% of the future reserve growth from discovered oil and gas fields in the Asia–Pacific region comes from Indonesia, in which the additional gas reserves are more than two times the additional oil reserves. Malaysia and China have similar potential for reserve growth in the future. The future reserve growth in these two countries account for 15.3 and 14.5% of the total future reserve growth in the Asia–Pacific region (Figs. 11.6 and 11.7).

The future reserve growth in discovered oil and gas fields in Indonesia are 26.4×10^8 tons of oil equivalent, of which oil accounts for 29.6%, condensate accounts for 4.6%, and gas accounts for 65.8%. The future reserve growth in Malaysia are 16.5×10^8 tons of oil equivalent, of which oil accounts for 9.6%, condensate accounts for 4.0%, and gas accounts for 86.4%. The future reserve growth in China are 15.8×10^8 tons of oil equivalent, of which oil accounts for 30.8%, condensate accounts for 2.2%, and gas accounts for 67.0%.

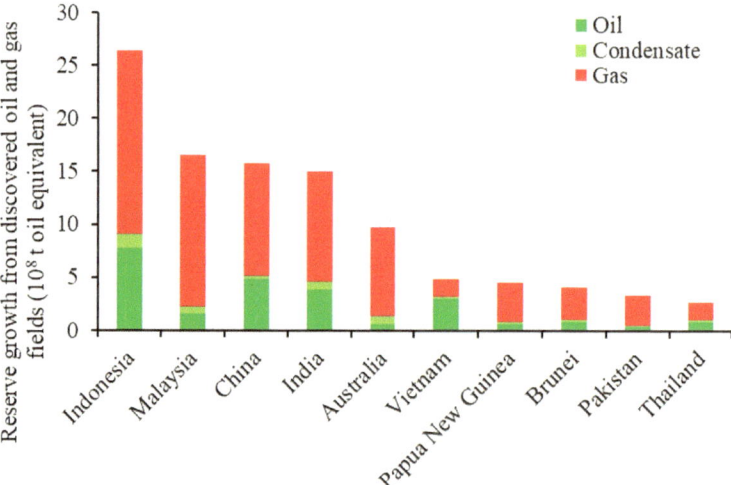

Fig. 11.6 Histogram of future reserve growth in discovered oil and gas fields of major countries in the Asia–Pacific region

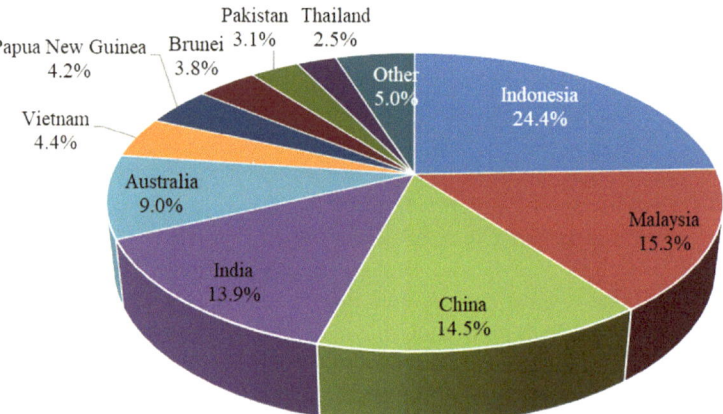

Fig. 11.7 Pie chart of future reserve growth in discovered oil and gas fields of major countries in the Asia–Pacific region

11.1.2.2   Distribution of Reserve Growth in Discovered Oil and Gas Fields in Major Basins

The future reserve growth in the Asia–Pacific region are mainly distributed in 68 basins, including the Zengmu Basin, Brunei-Sabah Basin, Krishna-Godavari Basin, Kutei Basin, and North Carnarvon Basin. Among these basins, the Zengmu Basin, Brunei-Sabah Basin and Krishna-Godavari Basin rank among the top three in terms of future reserve growth. The future reserve growth in the Asia–Pacific region are evenly distributed. The sum of reserve growth in discovered oil and gas fields in the top ten basins accounts for only about 52.7% of the total future reserve growth in the Asia–Pacific region (Figs. 11.8 and 11.9).

The future reserve growth in discovered oil and gas fields in the Zengmu Basin are 9.3×10^8 tons of oil equivalent, of which oil accounts for 7.8%, condensate

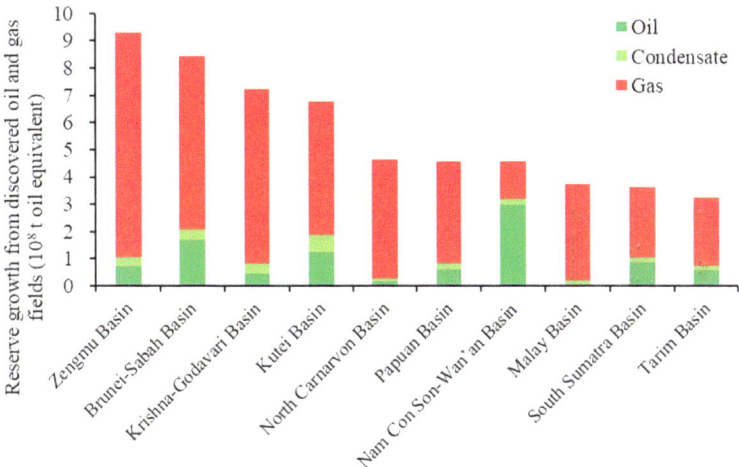

Fig. 11.8 Histogram of future reserve growth in discovered oil and gas fields of major basins in the Asia–Pacific region

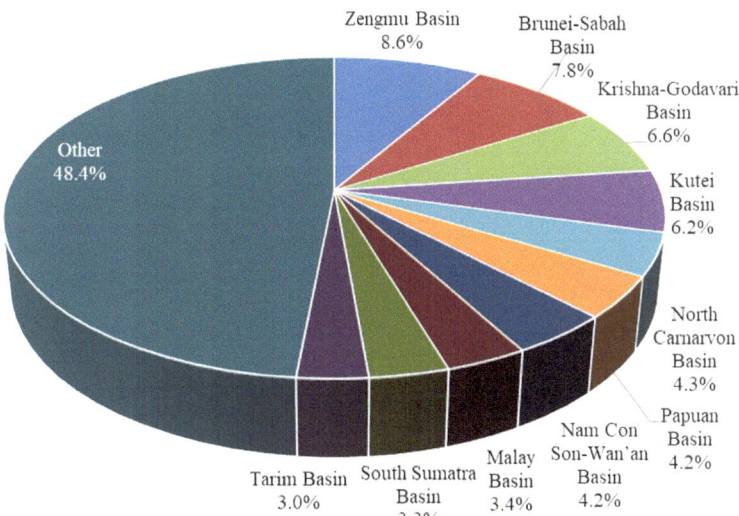

Fig. 11.9 Pie chart of future reserve growth in discovered oil and gas fields in major basins in the Asia–Pacific region

accounts for 3.4%, and gas accounts for 88.8%. The future reserve growth in the Brunei-Sabah Basin are 8.4×10^8 tons of oil equivalent, of which oil accounts for 20.1%, condensate accounts for 4.5%, and gas accounts for 75.4%. The future reserve growth in the Krishna-Godavari Basin are 7.2×10^8 tons of oil equivalent, of which oil accounts for 6.5%, condensate accounts for 5.2%, and gas accounts for 88.3%.

11.1.2.3 Distribution of Reserve Growth in Discovered Onshore and Offshore Oil and Gas Fields

In the Asia–Pacific region, the future reserve growth in discovered offshore oil and gas fields account for 64.1% of the total future reserve growth in this region. In both onshore and offshore areas, the future additional gas reserves are significantly more than the additional oil reserves (Fig. 11.10).

The future reserve growth in discovered offshore oil and gas fields in the Asia–Pacific region are 69.5×10^8 tons of oil equivalent, of which oil accounts for 27.9%, condensate accounts for 6.1%, and gas accounts for 66%. The future reserve growth in discovered onshore oil and gas fields in the Asia–Pacific region are 38.9×10^8 tons of oil equivalent, of which oil accounts for 45.1%, condensate accounts for 3.2%, and gas accounts for 51.7%.

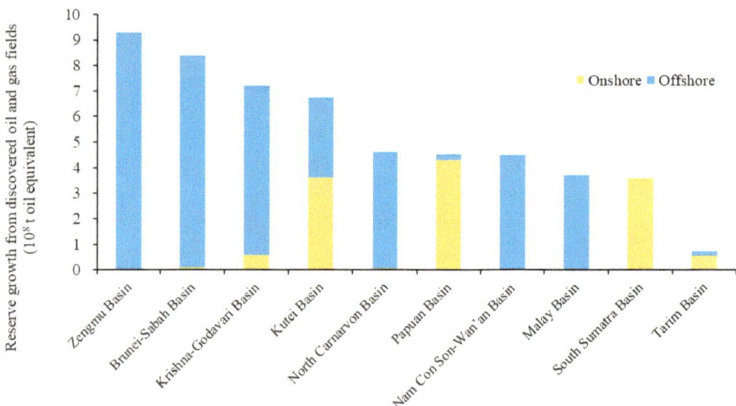

Fig. 11.10 Histogram of future reserve growth in the discovered onshore and offshore fields in major basins of the Asia–Pacific region

11.1.3 Characteristics of Distributions of Undiscovered Oil and Gas Resources

The undiscovered oil and gas resources in the Asia–Pacific region amount to 676.3 $\times$ 10^8 tons of oil equivalent, which are mainly gas and include 221.1 $\times$ 10^8 tons of oil, 21.7 $\times$ 10^8 tons of condensate, and 50.7 $\times$ 10^{12} m^3 of gas.

11.1.3.1 Distribution of Undiscovered Oil and Gas Resources in Major Countries

China has the largest amounts of undiscovered oil and gas resources, which account for 72.7% of the total undiscovered oil and gas resources in the Asia–Pacific region; and the undiscovered oil and gas resources in Australia and Indonesia account for 5.7% and 4.5% of the total undiscovered oil and gas resources in the Asia–Pacific region, respectively (Figs. 11.11 and 11.12).

The undiscovered recoverable oil and gas resources in China are 491.4 $\times$ 10^8 tons of oil equivalent, of which oil accounts for 32.6%, and gas accounts for 67.4%. The undiscovered recoverable oil and gas resources in Australia are 38.4 $\times$ 10^8 tons of oil equivalent, of which oil accounts for 41.9%, condensate accounts for 13.2%, and gas accounts for 44.9%. The undiscovered recoverable oil and gas resources in Indonesia are 30.3 $\times$ 10^8 tons of oil equivalent, of which oil accounts for 34.5%, condensate accounts for 18.2%, and gas accounts for 47.3%.

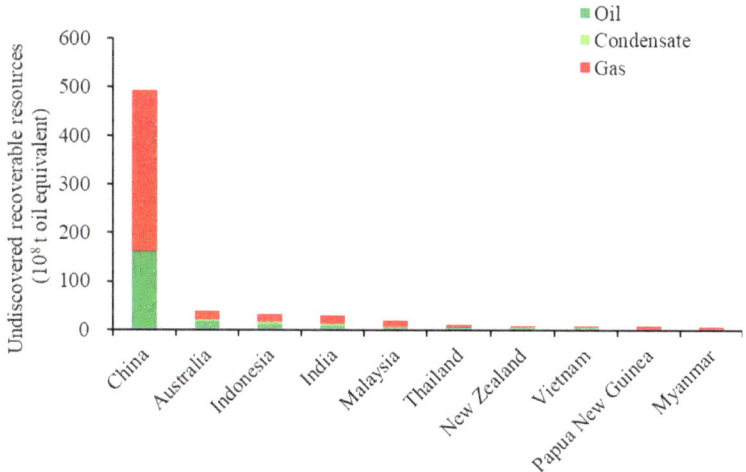

Fig. 11.11 Histogram of undiscovered recoverable oil and gas resources in major countries of the Asia–Pacific region

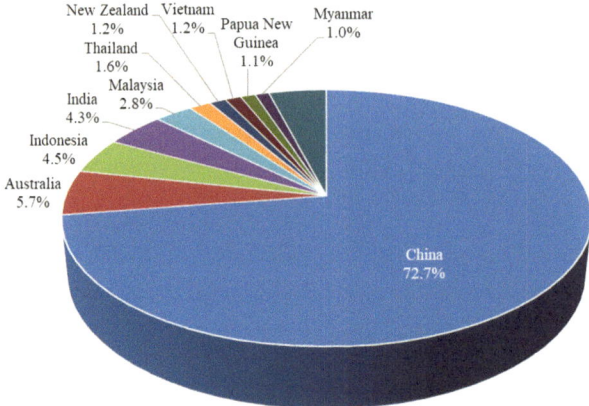

Fig. 11.12 Pie chart of undiscovered recoverable oil and gas resources in major countries of the Asia–Pacific region

11.1.3.2   Distribution of Undiscovered Oil and Gas Resources in Major Basins

The undiscovered oil and gas resources in the Asia–Pacific region are distributed in 105 basins. The undiscovered oil and gas resources in the top ten basins, including the Tarim Basin, Bohai Bay Basin, Sichuan Basin, Zengmu Basin, Pearl River Mouth Basin, and others, account for 53% of the total undiscovered oil and gas resources in the Asia–Pacific region (Figs. 11.13 and 11.14).

The undiscovered oil and gas resources in the Tarim Basin are 63×10^8 tons of oil equivalent, of which oil accounts for 24.5%, and gas accounts for 75.5%. The undiscovered oil and gas resources in the Bohai Bay Basin are 57.4×10^8 tons of oil equivalent, of which oil accounts for 76.2%, and gas accounts for 23.8%. The undiscovered oil and gas resources in Sichuan Basin are 50.9×10^8 tons of oil equivalent, mainly nature gas.

11.1.3.3   Distribution of Undiscovered Oil and Gas Resources in Onshore and Offshore Areas

The undiscovered oil and gas resources in the offshore areas of the Asia–Pacific region are more than those in the onshore areas. The undiscovered oil and gas resources in offshore and onshore areas account for 57.6% and 42.4%, respectively. In offshore areas in the Asia–Pacific region, the undiscovered oil resources are close to the undiscovered gas resources. In onshore areas, the undiscovered gas resources are slightly more than the undiscovered oil resources (Fig. 11.15).

The undiscovered recoverable oil and gas resources in offshore area are 389.5 $\times 10^8$ tons of oil equivalent, of which oil accounts for 27.8%, condensate accounts for 11.4%, and gas accounts for 60.8%. The undiscovered recoverable oil and gas

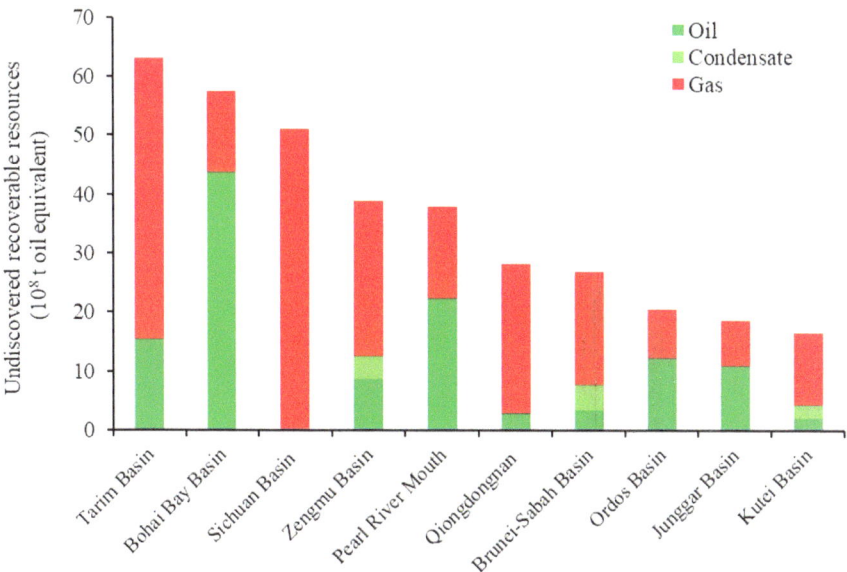

Fig. 11.13 Histogram of undiscovered recoverable oil and gas resources in major basins of the Asia–Pacific region

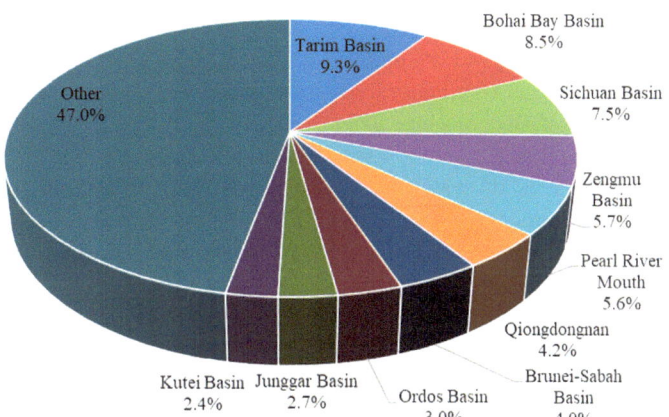

Fig. 11.14 Pie chart of undiscovered recoverable oil and gas resources in major basins in the Asia–Pacific region

resources in onshore area are 286.8×10^8 tons of oil equivalent, of which oil accounts for 36.3%, condensate accounts for 9.8%, and gas accounts for 53.9%.

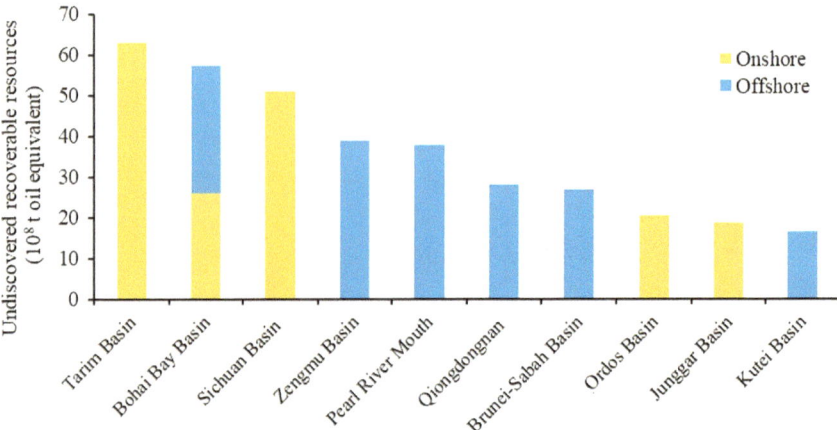

Fig. 11.15 Histogram of undiscovered recoverable oil and gas resources in onshore and offshore areas of major basins in the Asia–Pacific region

11.2 Unconventional Oil and Gas Resources

The unconventional oil and gas resources in the Asia–Pacific region mainly include oil shale, heavy oil, shale oil, tight gas, and CBM. The recoverable unconventional oil and gas resources in this region amount to 872.5×10^8 tons of oil equivalent, accounting for 12% of the world's total recoverable unconventional resources.

The recoverable unconventional oil resources in the Asia–Pacific region amount to 343.3×10^8 tons of oil equivalent, accounting for 8.1% of the world's total recoverable unconventional oil resources. These recoverable unconventional oil resources include 68.8×10^8 tons of heavy oil, accounting for 20% of the total recoverable unconventional oil resources in the Asia–Pacific region, 94.6×10^8 tons of shale oil, accounting for 27.6%, and 172.4×10^8 tons of oil shale, accounting for 50.2% (Figs. 11.16 and 11.17).

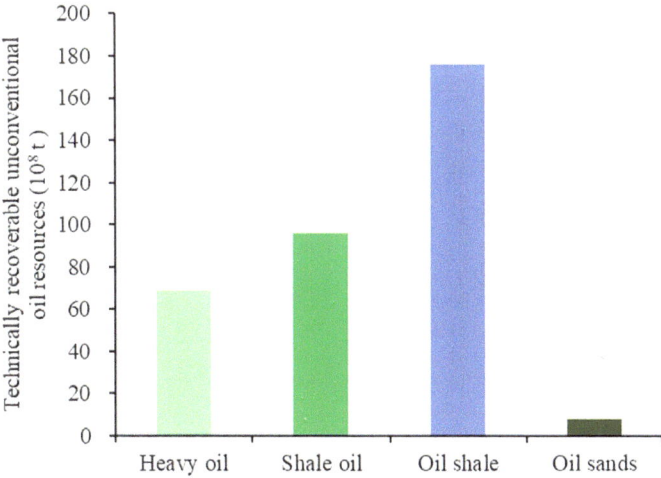

Fig. 11.16 Histogram of technically recoverable unconventional oil resources in the Asia–Pacific region

Fig. 11.17 Pie chart of technically recoverable unconventional oil resources in the Asia–Pacific region

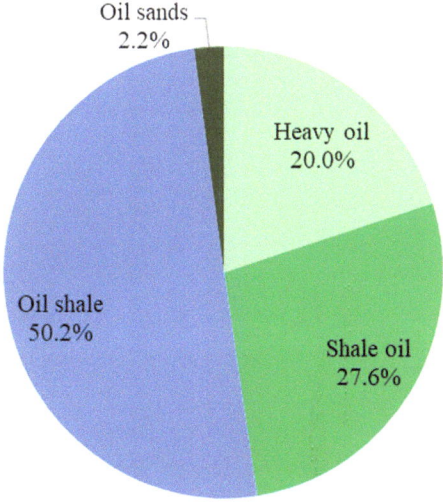

The recoverable unconventional gas resources in the Asia–Pacific region amount to 61.9×10^{12} m^3, accounting for 17.6% of the world's total unconventional gas resources. These unconventional gas resources include 34.3×10^{12} m^3 of shale gas, accounting for 55.4% of the total recoverable unconventional gas resources in the Asia–Pacific region, 17.1×10^{12} m^3 of CBM, accounting for 27.7%, and 10.5×10^{12} m^3 of tight gas (Figs. 11.18 and 11.19).

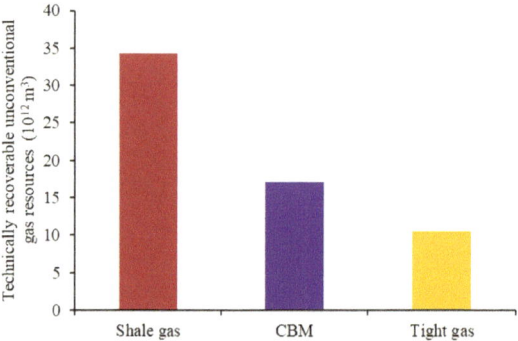

Fig. 11.18 Histogram of technically recoverable unconventional gas resources in the Asia–Pacific region

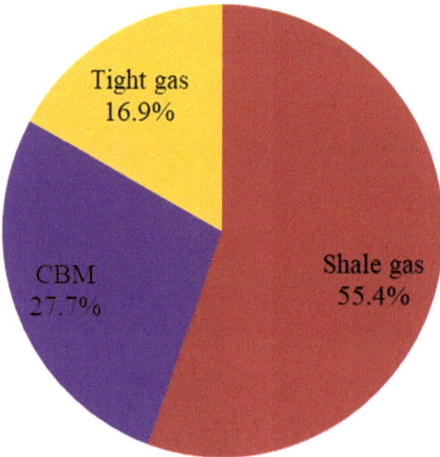

Fig. 11.19 Pie chart of technically recoverable unconventional gas resources in the Asia–Pacific region

11.2.1   *Distribution of Recoverable Unconventional Oil and Gas Resources in Major Countries (Regions)*

11.2.1.1   Distribution of Recoverable Unconventional Oil Resources in Major Countries (Regions)

The unconventional oil resources in the Asia–Pacific region are mainly distributed in China, Indonesia, Australia, India, and Malaysia. The recoverable unconventional oil resources in China amount to 198.8×10^8 tons, accounting for 57.9% of the total recoverable unconventional oil resources in the Asia–Pacific region; the recoverable unconventional oil resources in Indonesia are 60.3×10^8 tons, accounting for 17.6%;

and the recoverable unconventional oil resources in Australia are 48.1×10^8 tons, accounting for 14% (Figs. 11.20 and 11.21).

The recoverable unconventional oil resources in China mainly include oil shale and shale oil, which account for 69.9% and 26.4%, respectively. The recoverable unconventional oil resources in China mainly include 138.9×10^8 tons of oil shale and 52.4×10^8 tons of shale oil. Heavy oil is the major type of unconventional oil

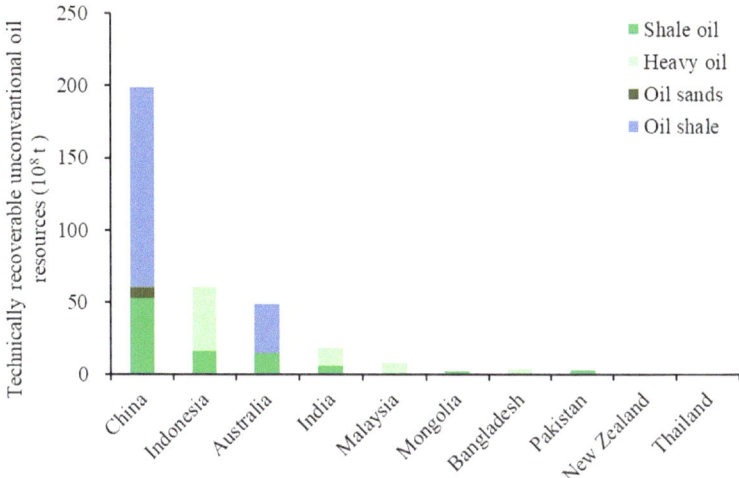

Fig. 11.20 Histogram of technically recoverable unconventional oil resources distributed in major countries in the Asia–Pacific region

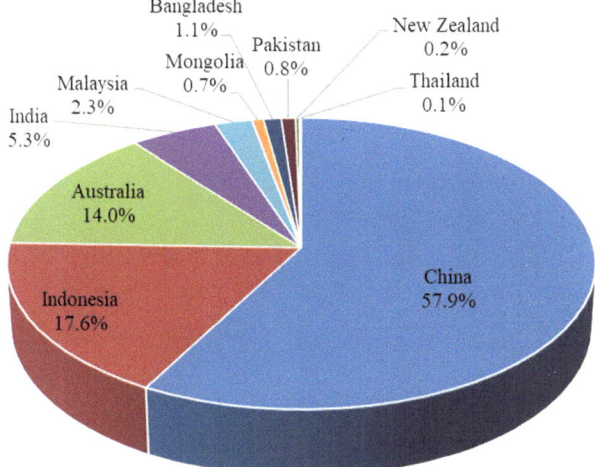

Fig. 11.21 Pie chart of technically recoverable unconventional oil resources distributed in major countries in the Asia–Pacific region

resource in Indonesia, and the recoverable heavy oil resources in Indonesia account for 74.3% of the total unconventional oil resources in this country, and the rest is shale oil.

11.2.1.2   Distribution of Recoverable Unconventional Gas Resources in Major Countries (Regions)

The unconventional gas resources in the Asia–Pacific region are mainly concentrated in China, Australia, Indonesia, India, and Pakistan. The recoverable unconventional gas resources in China are 33.5×10^{12} m^3, accounting for 54.1% of the total recoverable unconventional gas resources in the Asia–Pacific region and including 12.1 $\times$ 10^{12} m^3 of shale gas, 11.2×10^{12} m^3 of CBM and 10.3×10^{12} m^3 of tight gas. The recoverable unconventional gas resources in Australia are 15.1×10^{12} m^3, accounting for 24.4% of the total recoverable unconventional gas resources in the Asia–Pacific region and including 11.7×10^{12} m^3 of shale gas and 3.3×10^{12} m^3 of CBM. The recoverable unconventional gas resources in Indonesia are 6.6×10^{12} m^3, accounting for 10.6% of the total recoverable unconventional gas resources in the Asia–Pacific region and including 4.6×10^{12} m^3 of shale gas and 2.0×10^{12} m^3 of CBM. The recoverable unconventional gas resources in India are 3.4×10^{12} m^3, of which shale gas resources account for 79.9% (Figs. 11.22 and 11.23).

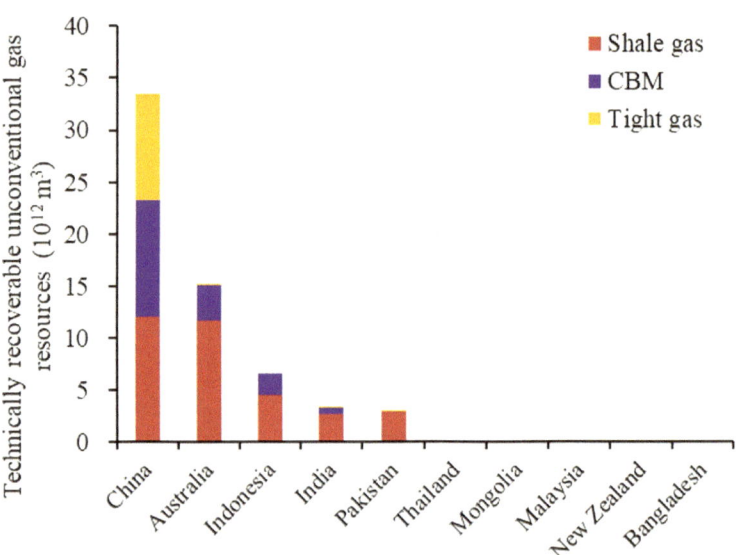

Fig. 11.22 Histogram of technically recoverable unconventional gas resources in major countries of the Asia–Pacific region

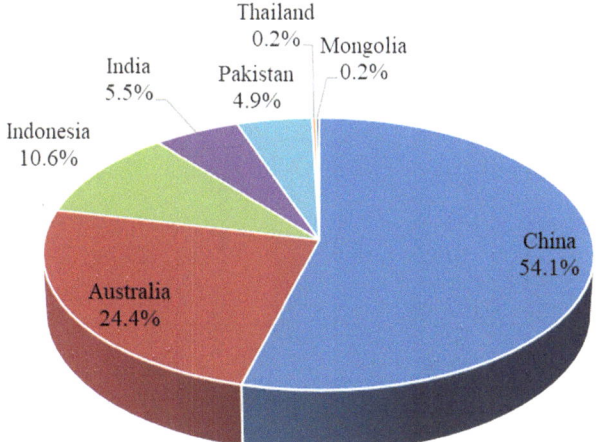

Fig. 11.23 Pie chart of technically recoverable unconventional gas resources in major countries of the Asia–Pacific region

11.2.2 Distribution of Unconventional Oil and Gas Resources in Major Basins

The unconventional oil and gas resources in the Asia–Pacific region are unevenly distributed in various basins. 90.0% of the unconventional oil and gas resources in this region are concentrated in the top ten basins.

11.2.2.1 Distribution of Unconventional Oil Resources in Major Basins

The unconventional oil and gas resources in the Asia–Pacific region are mainly distributed in 34 basins. Among the basins in the Asia–Pacific region, the Ordos Basin in China ranks first in terms of recoverable unconventional oil resources. The recoverable unconventional oil resources in this basin are 77.8×10^8 tons, which account for 22.7% of the total recoverable unconventional oil resources in the Asia–Pacific region and mainly include heavy oil. The Songliao Basin ranks second, and the recoverable unconventional oil resources in this basin are 69.5×10^8 tons, accounting for 20.2% of the total recoverable unconventional oil resources in the Asia–Pacific region and mainly including oil shale and shale oil. The Central Sumatra Basin ranks third, and the recoverable unconventional oil resources in this basin are 38.6×10^8 tons, accounting for 11.3% of the total recoverable unconventional oil resources in the Asia–Pacific region and mainly including heavy oil and shale oil (Figs. 11.24 and 11.25).

Three-fourths of the recoverable shale oil resources in the Asia–Pacific region are concentrated in the Ordos Basin, Songliao Basin, South Sumatra Basin, Central Sumatra Basin, North Sumatra Basin, Canning Basin, McArthur Basin, Cambay

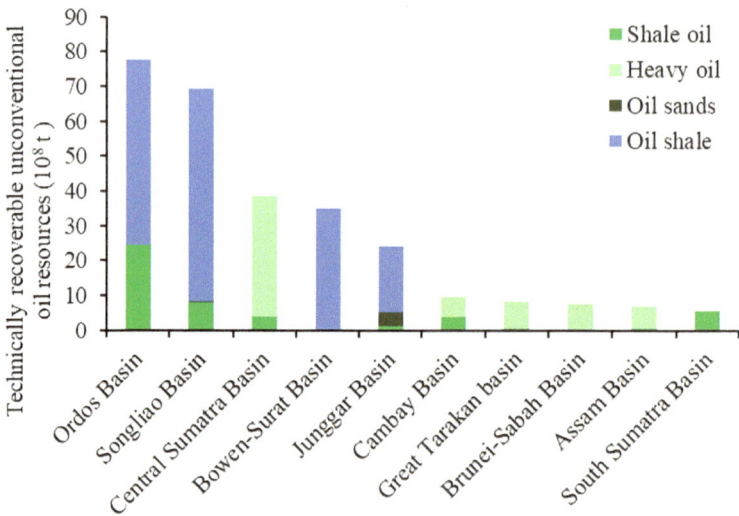

Fig. 11.24 Histogram of technically recoverable unconventional oil resources in major basins in the Asia–Pacific region

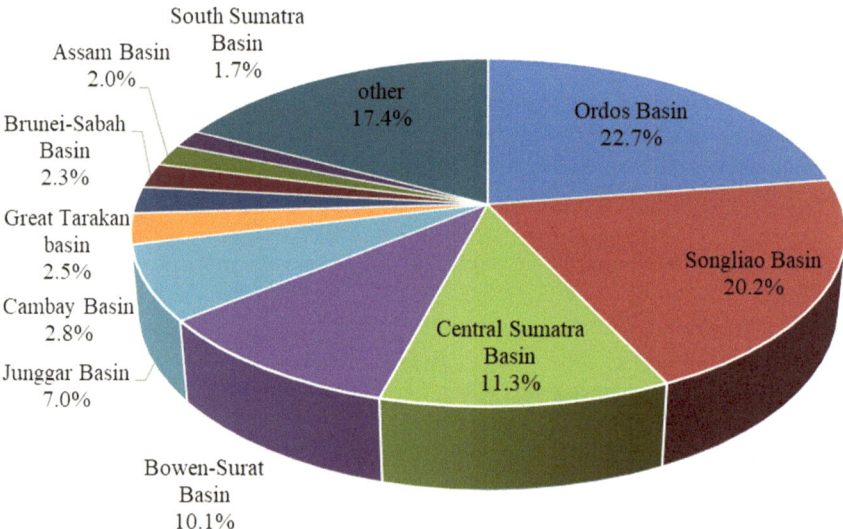

Fig. 11.25 Pie chart of technically recoverable unconventional oil resources in major basins in the Asia–Pacific region

Basin, Eromanga Basin, and Indus River Basin. About half of the recoverable heavy oil resources in this region are concentrated in the Central Sumatra Basin. All the recoverable oil sands in the Asia–Pacific region are concentrated in the Junggar Basin. The recoverable oil shale in this region are distributed in Ordos Basin and Songliao Basin.

11.2.2.2   Distribution of Unconventional Gas Resources in Major Basins

The unconventional gas resources in the Asia–Pacific region are mainly distributed in 32 basins, among which the Sichuan Basin in China ranks first. The recoverable unconventional gas resources in the Basin amount to 11.6×10^{12} m^3, accounting for 18.7% of the total recoverable unconventional gas resources in the Asia–Pacific region and mainly including shale gas. The Ordos Basin ranks second, and the recoverable unconventional gas resources in this basin are 10.8×10^{12} m^3, accounting for 17.4% of the total recoverable unconventional gas resources in the Asia–Pacific region and mainly consisting of tight gas. The Canning Basin ranks third, and the recoverable unconventional gas resources in this basin are 6.6×10^{12} m^3, accounting for 10.7% of the total recoverable unconventional gas resources in the Asia–Pacific region (Figs. 11.26 and 11.27).

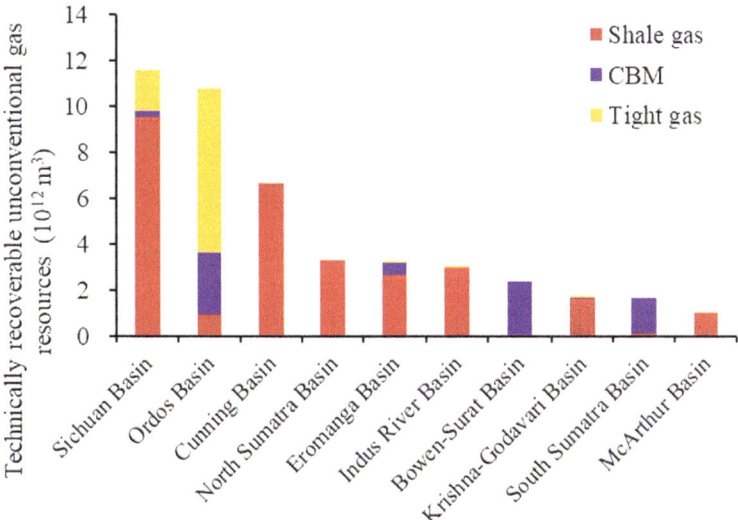

Fig. 11.26 Histogram of technically recoverable unconventional gas resources distributed in major basins in the Asia–Pacific region

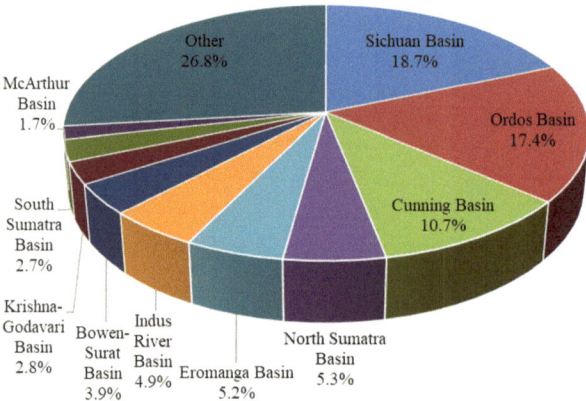

Fig. 11.27 Pie chart of technically recoverable unconventional gas resources distributed in major basins in the Asia–Pacific region

In the Asia–Pacific region, 77.3% of recoverable shale gas resources are concentrated in the Sichuan Basin, Canning Basin, North Sumatra Basin, Indus River Basin, Eromanga Basin, and Krishna-Godavari Basin; 64.8% of recoverable CBM resources are concentrated in the Bowen-Surat Basin and South Sumatra Basin; and all tight gas resources are distributed in the Ordos Basin, Eromanga Basin, Krishna-Godavari Basin, Perth Basin, Cambay Basin, and Indus River Basin.

Chapter 12
Key Exploration Areas and Cooperation Directions in the Future

12.1 Key Exploration Areas in the Future

Depending on the extent of exploration, the potential of undiscovered oil and gas resources has been evaluated using scientific methods based on a systematic analysis of the conditions for oil and gas accumulation in each basin. The evaluation results mainly represent the exploration potential of each basin, especially the potential and direction of exploration tasks involving risks. The potential of conventional and unconventional oil and gas resources and the key areas for exploration of such resources in the future can be identified by combining the geological conditions of oil and gas-bearing plays with the potential of undiscovered, recoverable conventional oil and gas resources and the potential of technically recoverable unconventional oil and gas resources.

© The Author(s) 2024
L. Dou et al., *Global Oil and Gas Resources: Potential and Distribution*,
https://doi.org/10.1007/978-981-97-4756-6_12

12.1.1   Key Exploration Areas for Conventional Oil Resources

The key exploration areas for conventional oil resources are mainly distributed in the following seven major groups of basins: (1) the basins along the passive continental margin of the North Atlantic, including the Scotia Basin, the Grand Banks, and the Tarfaya Basin, where the main plays are the Jurassic-Cretaceous fault-block structures and turbidite-sand body composite traps; (2) the basins along the passive continental margin of the Mid-Atlantic, including the portion of the Gulf of Mexico Basin in the USA, the Sureste Basin, and the basins in the periphery of the Caribbean Sea, where the main plays are the Cretaceous rift-system structural traps and the Paleogene-Neogene lithologic traps; (3) the basins along the passive continental margin of the northern part of the South Atlantic, including the Foz do Amazon Basin, the Suriname Basin, and Guyana coastal basins, where the main plays are the Cretaceous-Miocene slope fans and basin floor fans; (4) the basins along the passive continental margin of the middle part of the South Atlantic, including the Santos Basin, the Campos Basin, and the Kwanza Basin, where the main plays are supra-salt gravity flow depositional systems, pre-salt carbonates, and pre-salt sandstones; (5) the basins along the passive continental margin basin of the southern part of the South Atlantic, including the Colorado Basin and the Salado Basin, where the main plays are the Cretaceous rift-system structural and lithologic traps; (6) the Russian foreland-rift basins, where the main plays in the large rift basins in West Siberia are the structural traps in the northern offshore areas and the Upper Jurassic lithostratigraphic traps in the southern onshore areas; the favorable oil and gas in the Timan-Pechora Basin are the structural traps in the Devonian formations; and the favorable oil and gas accumulation plays in the East Siberian Basin are the Riphean carbonate reservoirs and Vendian clastic reservoirs; and (7) the Zagros/Arabian foreland basins, where the main favorable plays are the Paleozoic, Mesozoic, Cenozoic anticlinal traps and complex thrust structures (Fig. 12.1, Table 12.1).

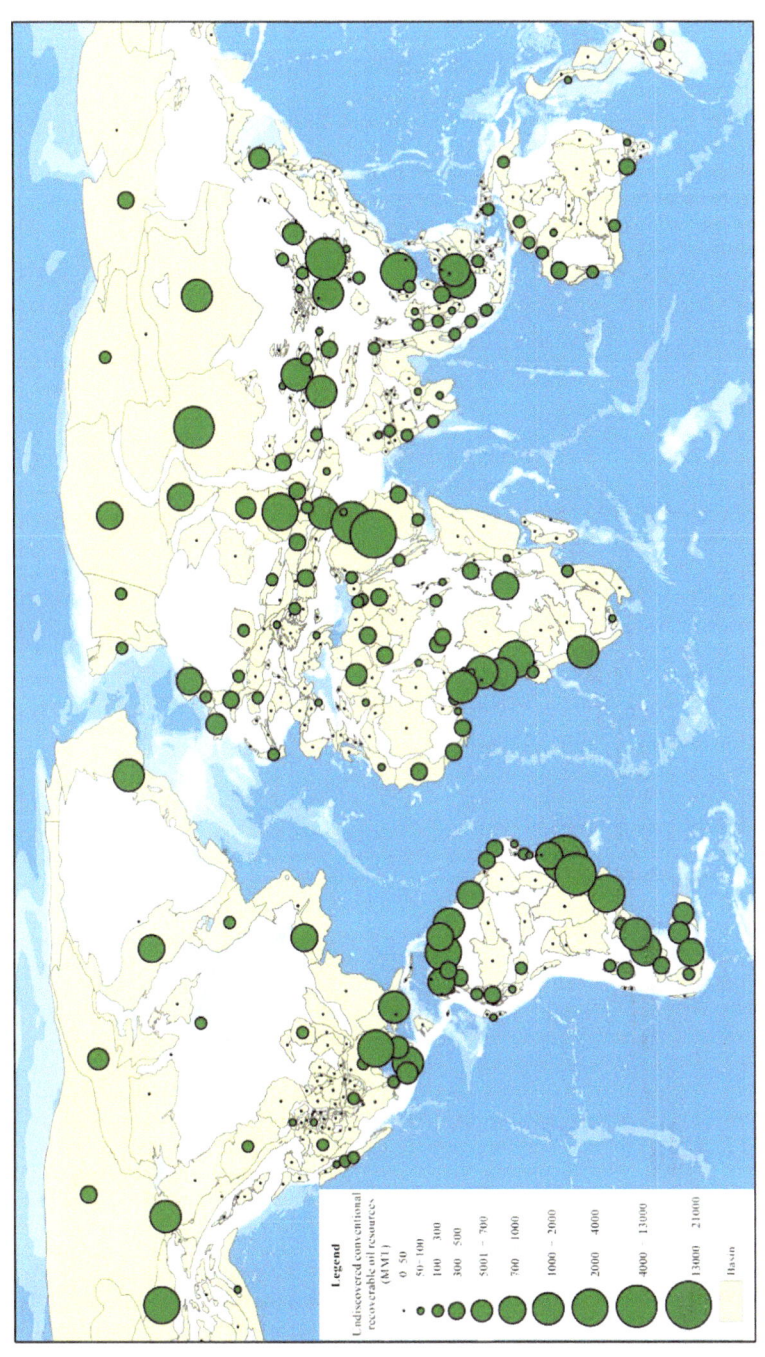

Fig. 12.1 Map showing the distribution of undiscovered recoverable conventional oil resources in various basins worldwide (Miller Cylindrical Projection)

Table 12.1 List of key exploration areas and directions for conventional oil resources

No.	Area/basin group	Major basins	Favorable plays
1	Basins along the passive continental margin of the North Atlantic	Scotia Basin, Grand Banks, Tarfaya Basin	Jurassic-Cretaceous fault-block structures and turbidite-sand body composite traps
2	Basins along the passive continental margin of the Mid-Atlantic	The portion of the Gulf of Mexico Basin in the USA, Sureste Basin, basins in the periphery of the Caribbean Sea	Cretaceous rift-system structural traps, Paleogene-Neogene lithologic traps
3	Basins along the passive continental margin of the northern part of the South Atlantic	Foz do Amazon Basin, Suriname Basin, Guyana coastal basins	Cretaceous-Miocene slope fans and basin floor fans
4	Basins along the passive continental margin of the middle part of the South Atlantic	Santos Basin, Campos Basin, Kwanza Basin	Supra-salt gravity flow depositional systems, pre-salt carbonates and sandstones
5	Basins along the passive continental margin of the southern part of the South Atlantic	Colorado Basin, Salado Basin	Cretaceous rift-system structural and lithologic traps
6	Russian foreland-rift basins	West Siberian Basin	Structural traps in the northern offshore areas and the Upper Jurassic lithostratigraphic traps in the southern onshore areas
		Timan-Pechora Basin	Structural traps in the Devonian formations
		East Siberian Basin	Riphean carbonate reservoirs, Vendian clastic reservoirs
7	Zagros/Arabian foreland basins	Zagros Basin, Arabian Basin	Paleozoic, Mesozoic, Cenozoic anticlinal traps and complex thrust structures

12.1.2   Key Exploration Areas for Conventional Gas Resources

The future exploration areas for conventional gas resources are mainly distributed in the following six major groups of basins: (1) the Zagros/Arabian foreland basins, mainly including the Zagros Basin and Arabian Basin, where the main plays are the large anticlinal traps in the Paleozoic and Mesozoic deep formations; (2) the passive margin basins in East African offshore, mainly including the Somali Basin, Zambezi

Delta Basin and Tanzania Coastal Basin, where the favorable plays are four major annular structural plays in the passive margin delta and the Cretaceous-Paleogene slope fans and basin floor fans; (3) the basins along the passive continental margin of the North Atlantic, including the Scotia Basin, the Grand Banks, and the Tarfaya Basin, where the favorable plays are the Jurassic-Cretaceous fault-block structures and turbidite-sand body composite traps; (4) the Central Asian rift-foreland basins, where the favorable plays include the Paleozoic (pre-salt) carbonate reservoirs in the Precaspian Basin, Chu-Sarysu Basin and Syr Darya Basin, the Paleogene-Neogene structural traps in the South Caspian Basin, and the Jurassic pre-salt bioherms and sandstone lithologic traps in the Amu Darya Basin; (5) the Russian foreland-rift basins, where the favorable plays include the large structural traps in the northern offshore areas of the West Siberian Basin and the Riphean carbonate reservoirs and Vendian clastic reservoirs in the East Siberian Basin; and (6) the Arctic passive margin basins, mainly including Kara Sea and Barents Sea basins, where the favorable plays are the Jurassic-Cretaceous fault-block structural traps (Fig. 12.2; Table 12.2).

12.1.3 Key Exploration Areas for Shale Oil Resources

The favorable exploration areas for shale oil resources are mainly distributed in the following seven major groups of basins: (1) the North American foreland basins, where the main plays in the Permian Basin, Appalachian Basin and Williston Basin are the Devonian-Carboniferous formations and those in the Gulf Coast Basin, Denver Basin and Powder River Basin are the Cretaceous strata; (2) the Andean foreland basins, mainly including the Llanos Basin, Putumayo Basin, Maranon Basin, and Neuquén Basin, where the favorable plays are the Jurassic-Cretaceous Formations; (3) the North African cratonic basins, mainly including the Atlas fold belt and Ghadames Basin, where the favorable plays are the Silurian and Devonian Formations; (4) the Central and West African rift basins, mainly including the Bongor Basin, Termit Basin, South Chad Basin, Muglad Basin, and Melut Basin, where the favorable plays are the Cretaceous lacustrine shale oil reservoirs; (5) the Russian foreland-rift basins, where the favorable play includes the Upper Jurassic Bazhenov Formation in the West Siberian Basin and the Devonian Domanic Formation in the Timan-Pechora Basin and Volga-Ural Basin; (6) the Zagros/Arabian foreland basins, mainly including the Zagros Basin and Arabian Basin, where the favorable plays are the Jurassic and Cretaceous marine shale oil reservoirs; and (7) the Southeast Asian back-arc basins, mainly including the North Sumatra Basin, Central Sumatra Basin and South Sumatra Basin, where the favorable plays are the Cenozoic lacustrine shale oil reservoirs (Fig. 12.3; Table 12.3).

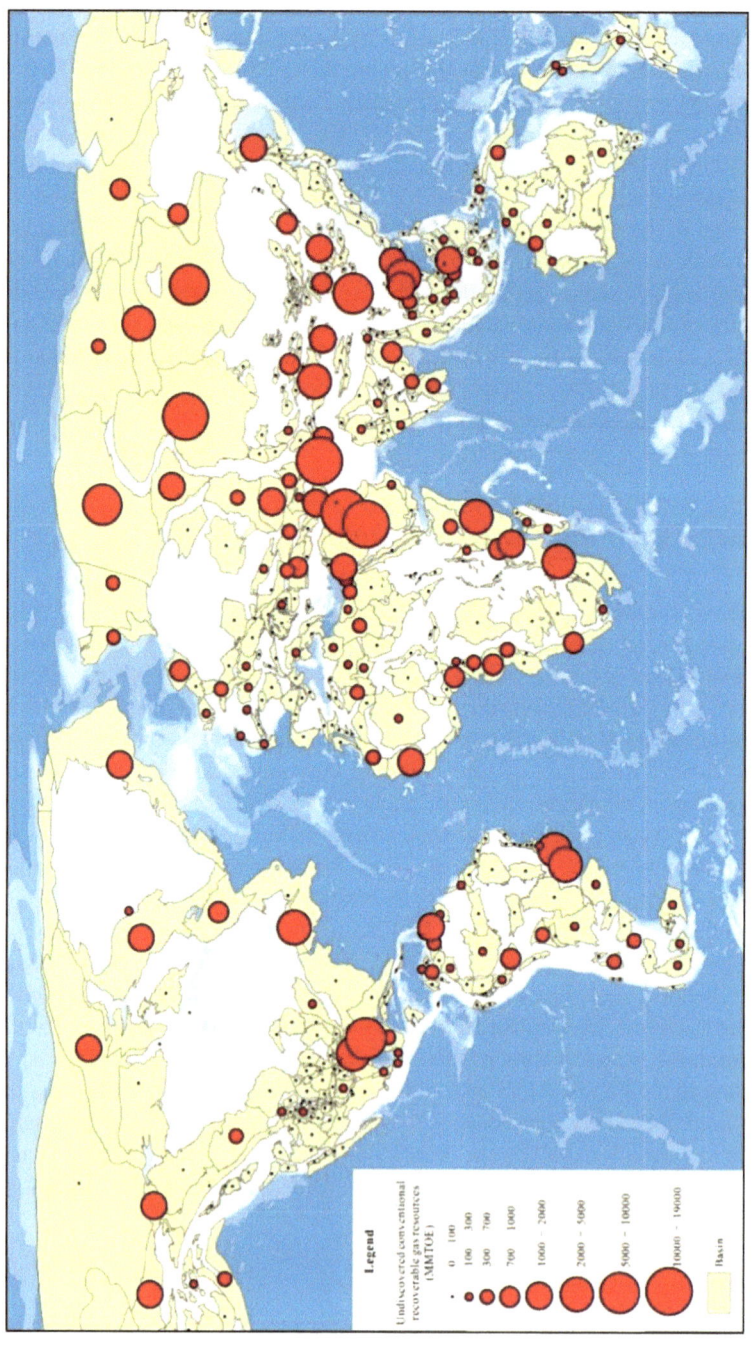

Fig. 12.2 Map showing the distribution of undiscovered recoverable conventional gas resources in various basins worldwide (Miller Cylindrical Projection)

Table 12.2 List of key exploration areas and directions for conventional gas resources

No.	Area/basin group	Major basins	Favorable oil and gas accumulation plays
1	Zagros/Arabian foreland basins	Zagros Basin, Arabian Basin	Large anticlinal traps in the Paleozoic and Mesozoic deep formations
2	Passive margin basins in East African offshore	Somali Basin, Zambezi Delta Basin, Tanzania Coastal Basin	Four major annular structural plays in the passive margin delta and the Cretaceous-Paleogene slope fans/basin floor fans
3	Basins along the passive continental margin of the North Atlantic	Scotia Basin, the Grand Banks, Tarfaya Basin	Jurassic-Cretaceous fault-block structures and turbidite-sand body composite traps
4	Central Asian rift-foreland basins	Precaspian Basin, Chu-Sarysu Basin, Syr Darya Basin	Paleozoic (pre-salt) carbonate reservoirs
		South Caspian Basin, Amu Darya Basin	Paleogene-Neogene structural traps, Jurassic pre-salt bioherms
5	Russian foreland-rift basins	West Siberian Basin	Large structural traps in offshore areas
		East Siberian Basin	Riphean carbonate reservoirs, Vendian clastic reservoirs
6	Arctic passive margin basins	Kara Sea and Barents Sea basins	Jurassic-Cretaceous fault-block structures

12.1.4 Key Exploration Areas for Shale Gas Resources

The key exploration areas for shale gas resources are mainly distributed in the following seven major groups of basins: (1) the North American foreland basins, where the favorable plays include the Devonian-Carboniferous formations in the Permian Basin, Appalachian Basin and Williston Basin and the Cretaceous formations in the Gulf Coast Basin and Powder River Basin; (2) the cratonic basins in central South America, mainly including the Chaco Basin, Cuyo Basin and Amazon Basin, where the favorable oil and gas accumulation plays are the Devonian-Carboniferous Formations; (3) the Andean foreland basins, mainly including the Neuquén Basin and Magallanes Basin, where the favorable plays are the Jurassic-Cretaceous Formations; (4) the North African cratonic basins, mainly including the Atlas fold belt and Ghadames Basin, where the favorable plays are the Silurian and Devonian Formations; (5) the Russian foreland-rift basins, where the favorable plays include the

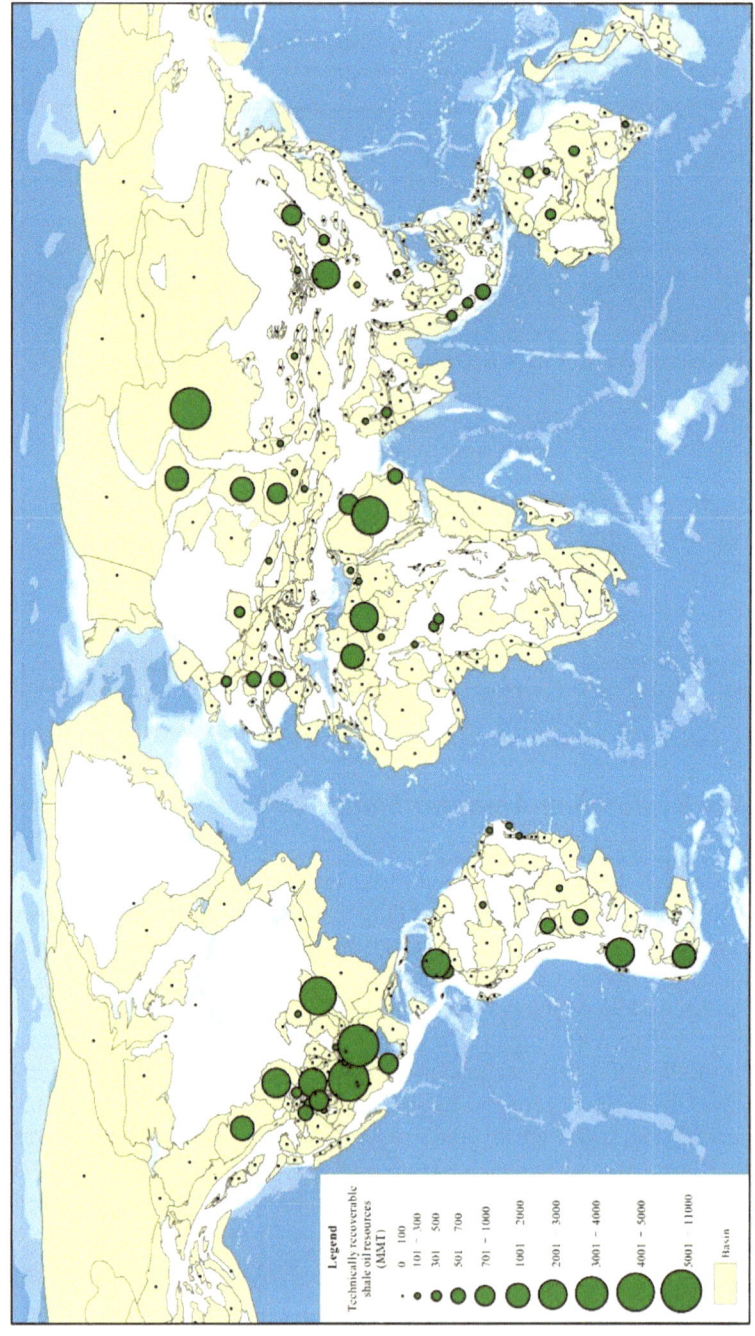

Fig. 12.3 Map showing the distribution of technically recoverable shale oil resources in various basins worldwide (Miller Cylindrical Projection)

Table 12.3 List of key exploration areas and directions for shale oil resources worldwide

No.	Area/basin group	Major basins	Favorable plays
1	North American foreland basins	Permian Basin, Appalachian Basin, Williston Basin	Devonian-Carboniferous
		Gulf Coast Basin, Denver Basin, Powder River Basin	Cretaceous
2	Andean foreland basins	Llanos Basin, Putumayo Basin, Maranon Basin, Neuquén Basin	Jurassic-Cretaceous
3	North African cratonic basins	Atlas fold belt, Ghadames Basin	Silurian and Devonian
4	Central and Western African rift basins	Bongor Basin, South Chad Basin, Muglad Basin, Melut Basin	Cretaceous shale oil reservoirs
5	Russian foreland-rift basins	West Siberian Basin	Upper Jurassic Bazhenov Formation
		Timan-Pechora Basin, Volga-Ural Basin	Devonian Domanic Formation
6	Zagros/Arabian foreland basins	Zagros Basin, Arabian Basin	Jurassic and Cretaceous marine shale oil reservoirs
7	Southeast Asian back-arc basins	North Sumatra Basin, Central Sumatra Basin, South Sumatra Basin	Cenozoic lacustrine shale oil reservoirs

Upper Jurassic Bazhenov Formation in the West Siberian Basin and the Devonian Domanic Formation in the Timan-Pechora Basin and Volga-Ural Basin; (6) the Zagros/Arabian foreland basins, mainly including the Zagros Basin and Arabian Basin, where the favorable plays are Jurassic and Cretaceous Formations; and (7) the craton basins in Central Australia, mainly including the Canning Basin, McArthur Basin and Eromanga Basin, where the favorable oil and gas accumulation plays are the Devonian-Carboniferous Formations (Fig. 12.4, Table 12.4).

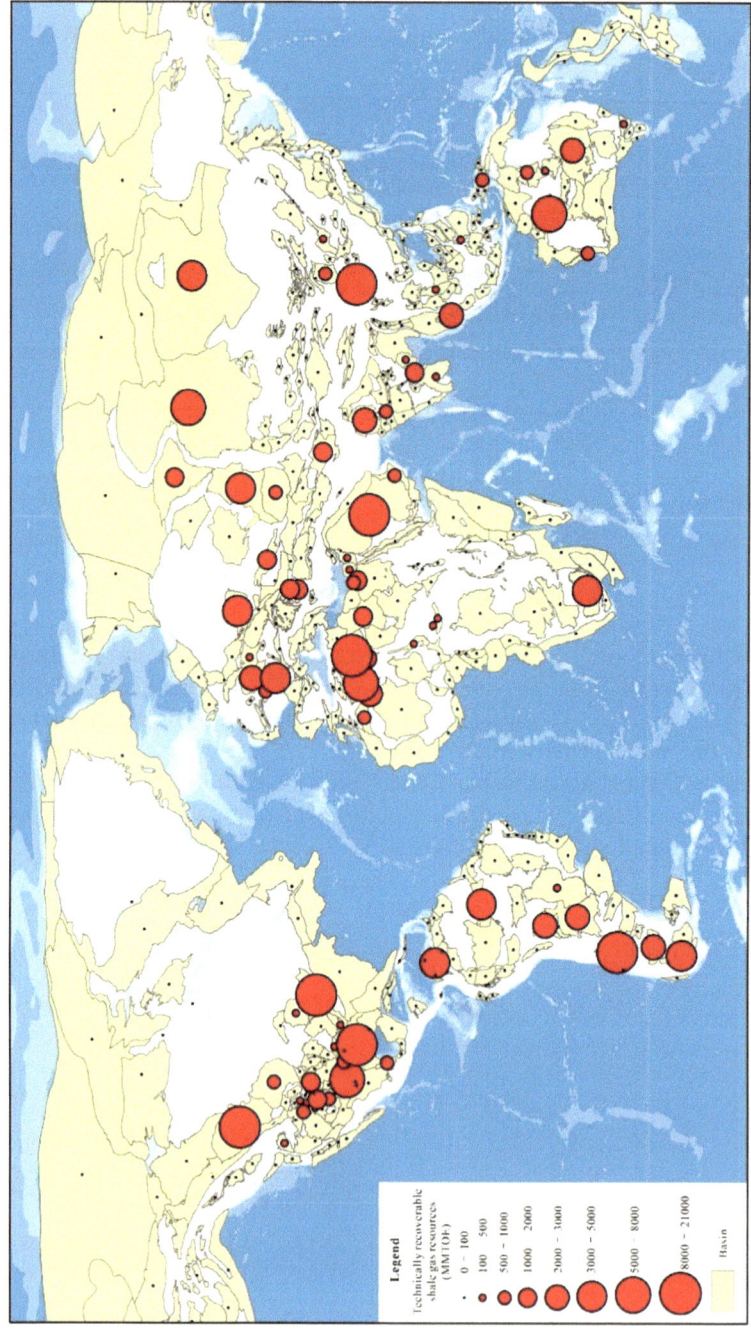

Fig. 12.4 Map showing the distribution of technically recoverable shale gas resources in various basins worldwide (Miller Cylindrical Projection)

Table 12.4 List of key exploration areas and directions for shale gas resources worldwide

No.	Area/basin group	Major basins	Favorable plays
1	North American foreland basins	Permian Basin, Appalachian Basin, Williston Basin	Devonian-Carboniferous
		Gulf Coast Basin, Denver Basin, Powder River Basin	Cretaceous formations
2	Cratonic basins in central and South America	Chaco Basin, Cuyo Basin, Amazon Basin	Devonian-Carboniferous
3	Andean foreland basins	Neuquén Basin, Magallanes Basin	Jurassic-Cretaceous
4	North African cratonic basins	Atlas fold belt, Ghadames Basin	Silurian and Devonian
5	Russian foreland-rift basins	West Siberian Basin	Upper Jurassic Bazhenov
		Timan-Pechora Basin, Volga-Ural Basin	Devonian Domanic
6	Zagros/Arabian foreland basins	Zagros Basin, Arabian Basin	Jurassic and Cretaceous
7	Craton basins in Central Australia	Canning Basin, McArthur Basin, Eromanga Basin	Devonian-Carboniferous

12.2 Cooperation Opportunities Worthy of Focused Attention in Future

Based on an analysis of the new oil and gas discoveries worldwide made through explorations in the past 10 years and considering the geological conditions of oil and gas-bearing plays, the potential of undiscovered oil and gas resources, and the environment for cooperation, it is recommended to focus on the opportunities for cooperation in Guyana offshore, the Caribbean offshore, both sides of the South Atlantic, South African offshore, the Argentine offshore, and other regions in the near future.

12.2.1 Coastal Basins in Guyana and Surrounding Areas

Since 2015, ExxonMobil has been consistently conducting oil exploration activities in the coastal basins of Guyana and has discovered 18 new oil fields, including the Liza Oilfield. Additionally, Tullow Oil Plc (Tullow) has made two new discoveries in the adjacent blocks. In addition, TotalEnergies made several new oil discoveries (including condensate discoveries and a few gas discoveries), such as Maka Central discovery in block 58 of the Suriname offshore in early 2020. The new discoveries mentioned above will continue to boost reserve growth. The confirmed oil and gas accumulations in these areas are mainly found in the Upper Cretaceous turbidite sand bodies in passive margin basins, but they also include the Miocene turbidite sand bodies. Only one oilfield, the Ranger Oilfield, has been discovered in the Upper Jurassic-Lower Cretaceous biohermic play. However, its exploration potential warrants further investigation.

12.2.2 Periphery of the Caribbean Offshore

Since 2019, Barbados, Saint Vincent, Cuba, and other countries around the Caribbean Sea have been initiating tender procedures for offshore exploration blocks. BHP Group Limited will further evaluate the exploration potential of the Pliocene turbidite sand bodies located in the ultra-deep Trinidad-Tobago offshore, building on its successful exploration in 2019. Seismic data from multi-user companies such as MCG and CGG reveals various types of potential traps that have developed on the periphery of the Caribbean Sea, including large anticlines in different basins in Barbados and Togo, as well as Miocene channels exhibiting seismic reflection anomalies. In general, the level of exploration in this region is relatively low. The geological structures are complex, the conditions for oil and gas accumulation vary greatly in different basins, and the principles of oil and gas accumulation are still unclear. With the continuous collection of multi-user data and the deepening understanding of the regional geological setting, this region will emerge as a crucial new area for exploration in the future.

12.2.3 Pre-salt Formations on Both Sides of the South Atlantic

With the conclusion of the Brazil 4th, 5th and 6th PSC (production sharing contracts) Pre-Salt Bid Rounds for the period from 2018 to 2019, the winning bidders will successively carry out exploratory drilling of pre-salt formations in the Santos Basin in accordance with the PSCs. ExxonMobil has performed exploratory drilling in the Uirapuru Block, which was awarded in the 4th pre-salt bid round and made an oil discovery in the pre-salt target interval. This oil discovery is named the Araucaria Oilfield. Petrobras and PetroChina have confirmed through drilling that the Aram block has over 1 billion tons of oil in place and the reserves are expected to continue increasing through drilling in the future. On the east coast of the Atlantic Ocean, TotalEnergies will set a new record for deepwater drilling in offshore areas once again. It plans to conduct exploratory drilling of the target interval, which is a pre-salt carbonate formation located in Block 48 of Angola, at a water depth of 3628 m. Therefore, the pre-salt oil and gas accumulation plays on both sides of the South

Atlantic are still an important cooperation opportunity that require focused attention in the near future.

12.2.4 South African Offshore

TotalEnergies has discovered two large gas condensate fields in the Outeniqua Basin in South African offshore and also find Venus giant oil field in Orange basin. These discoveries have confirmed the exploration potential of oil and gas accumulation plays composed of the Cretaceous gravity-flow sandstones and further boosted the confidence in exploration activities in South Africa, Namibian offshore, southern Mozambique, and the conjugate with Argentine offshore. Therefore, these fields are expected to become exploration hotspots for some time in the future. Many oil companies, including Shell, Tullow and Serica, have completed the deployment of exploration assets in the Namibian offshore and identified more than 50 traps to be drilled, including many types of plays, such as bioherms and turbidite sand bodies. The new round of tendering for exploration blocks in the offshore of Mozambique and the second round of tendering for exploration blocks in the offshore of Argentina will also attract more oil companies to actively participate in these tendering rounds.

12.2.5 The Argentine Offshore

The Argentine offshore are over 3000 km long from north to south, with seven oil and gas basins developing from north to south, including Colorado, the Argentine deepwater Basin, and the Malvinas. The total sedimentary area of these basins is approximately 1.2 million square kilometers. The Argentine offshore have been not open for a long time and the level of exploration is extremely low. In April 2019, the Argentine government implemented the first offshore bid round, which was attracted almost all the international major companies such as Shell, TotalEnergies, and ExxonMobil, and other giants. The BGP's multi-client seismic data reveals that the shallow Colorado Basin has underwent two stages of rifts and various types of reservoir combinations such as anticlines and fault blocks are developed; TGS's multi-client sesmic data shows that the deepwater area development in the Argentine offshore is similar to the Lower Cretaceous basin bottom fan system in western South Africa. Inspired by the discovery of the Venus oil field in the Orange Basin of South Africa, the argentinian offshore are conjugate with the Orange Basin, indicating the development of similar high-quality source reservoir combinations. Deepwater and shallow water oil and gas exploration in this region are highly attractive.

12.2.6   Northwest African Offshore

Since 2015, two types of plays, namely, SNE and Fan, have been discovered in the offshore of Senegal. Several oil companies, such as Kosmos, BP and TotalEnergies, have quickly followed up and acquired almost all the exploration blocks in the surrounding area. Through exploratory drilling, these oil companies have verified the existence of two important types of plays, namely, channel complexes and basin floor fans, under the background of carbonate platforms in Northwest African offshore. These basins extend from the northern offshore of Liberia in the south and extend northwards to the Moroccan offshore. It is foreseeable that many basins in Northwest African offshore and the eastern seas of the USA will become the key areas where oil companies search for these two types of plays. BP has already deployed exploration assets in these offshore and plans to build a new LNG supply base there. In these areas, there will be many cooperation opportunities worthy attention in the future.

12.2.7   East African Offshore Basins

The East African offshore basins were formed during the breakup of the Gondwana continent, undergoing two major stages: the Permian Jurassic rift and the Cretaceous Paleogene drift, forming a typical passive continental margin basin. From north to south, they are Somalia Coastal Basin, Lamu Basin, Tanzania Coastal Basin, Ruvuma Basin, Angoche basin and Zambezi Delta Basin, with a total area of over 3.7 million square kilometers.

In 2010, Anadarko, Eni and other companies began deep-water drilling and discovered a total of 28 large and medium-sized gas fields in the Luwuma and the Tanzanian Coastal Basin. The proven natural gas reserves were about 5.3 trillion cubic meters (S&P international, 2023), and the success rate of exploration wells was 67%. Based on existing findings, and multi-client sesmic data, research suggests that the factors for natural gas accumulation in deep-water areas of East Africa include a large sedimentary filling thickness (7000 to 14,000 m) in the basin, potential development of Triassic, Jurassic and Cretaceous source rocks, and the formation of lithological traps and superior reservoir performance in deep-water gravity flow sand bodies of the Upper Cretaceous and Paleogene. It is predicted that the recoverable natural gas resources to be discovered will be 10.2 trillion cubic meters. The exploration potential in the Somali and Mozambique offshore is the greatest among them. Analysis suggests that besides abundant natural gas resources, the possibility of oil generation cannot be ruled out, and exploration prospects are worth paying attention to and looking forward to.

12.2.8 Deep Plays in the Zagros Basin

Due to strong tectonic compression in the Zagros Basin, large and elongated NW–SE trending anticlinal traps have developed in this basin, and multiple thrust nappes are superimposed vertically. Early oil and gas exploration was mainly targeted at the Jurassic-Neogene shallow layers. Advances in related technologies, such as technologies for the acquisition and processing of seismic data from mountainous and steeply dipping strata and the modeling of complex structures, have made it possible to image the complex thrust systems in the Triassic and deeper formations in the Zagros Basin more accurately. New seismic data have revealed that many potential traps have developed in the deep formations of this basin and the main type of oil and gas resource is gas. As Iran and other countries attach more importance to clean energy such as gas, the exploration potential of gas in the deep formations of this basin will be released gradually.

12.2.9 Russian Arctic

Due to extreme cold related to the geographical location of the Russian Arctic and the limitations of technologies usable in this polar region, the extent of oil and gas exploration in the Russian Arctic is very low. Since 2014, major new gas discoveries have been made successively in the West Siberian Basin through exploration in the South Kara Sea within the Arctic Circle. The main plays in the offshore portion of this basin are consistent with those in the onshore portion. However, due to large burial depth of source rocks and high degree of thermal evolution, the major type of oil and gas resource is gas. Seismic data shows that there are still a large number of undrilled traps in the South Kara Sea, implying that the exploration potential of this area is huge. The successful implementation of the Yamal LNG Project, the Arctic LNG 2 Project, and other similar projects has made the idea of the "Ice Silk Road" a reality. Therefore, oil and gas exploration in this region will be further accelerated in the future.

12.2.10 East African Rift System

The East African Rift System consists of a series of intracontinental rift basins that began to form along the two branches (eastern and western branches) of the system in the Miocene. Six rift basins, including the Lake Albert Basin, Lake Tanganyika Basin and Lake Malawi Basin, have developed from north to south along the western branch, covering a total area of more than 11×10^4 km^2 across Uganda, the Democratic Republic of the Congo, Tanzania, and other countries and extending southwards to Mozambique. A series of small rift basins, such as the Turkana Basin, Lokichar

Basin and Lake Magadi Basin, have developed along the eastern branch, covering a total area of more than 9×10^4 km^2 across Kenya, Ethiopia, and Tanzania. By the end of 2020, 19 oil fields had been discovered on the east side of the Albert Rift located in the north section of the western branch (in Uganda), and 10 oil fields had been discovered in the Lokichar Basin located in the middle section of the eastern branch (in Kenya). The proven recoverable oil reserves in these two groups of oil fields amount to 2.36×10^8 t and 0.76×10^8 t, respectively. The success rates for drilling exploration wells in these fields are higher than 60%. As agreements for the construction of export pipelines are finalized and signed and the discovered oilfields are put into operation, the cooperation opportunities in other rift basins with similar conditions for oil and gas accumulation will be worthy of focused attention in the future.

Annex

Unit Conversion Table

1 mile (mile) = 1.609 km (kilometers)

1 m (meter) = 3.281 ft (feet) = 1.094 yd (yards)

1 km^2 (square kilometer) = 100 ha (hectares) = 247.1 acre (acres) = 0.386 $mile^2$ (square miles)

1 × 10^{12} ft^3 (trillion cubic feet) = 283.17 × 10^8 m^3 (one million cubic meters)

1 m^3 (cubic meter) = 1000 L (liters) = 35.315 ft^3 (cubic feet) = 6.29 bbl (barrels)

1 bbl (barrel) = 0.14 t (ton) (crude oil, global average)

1 × 10^{12} ft^3/d (trillion cubic feet per day) = 283.17 × 10^8 m^3/d (one million cubic meters per day) = 10.336 × 10^{12} m^3/a (trillion cubic meters per annum)

1 bbl/d (barrel per day) = 50 t/a (tons per annum) (crude oil, global average)

1 t (ton) = 7.3 bbl (barrels) (crude oil, global average)

1 bbl of crude oil = 5800 ft^3 of gas (calculated by average calorific value)

1 bbl of crude oil = 5.8 × 10^6 Btu (British thermal unit)

1 D (Darcy) = 1000 mD (millidarcy) = 1 μm^2 (square micron)

1 cm^2 (square centimeter) = 9.81 × 10^7 D

1 ft^3/bbl (cubic foot/barrel) = 0.2067 m^3/t (cubic meter/ton) (gas-oil ratio)

1 °F/100 ft = 1.8 °C/100 m (geothermal gradient)

1 t (ton) = 1000 kg (kilograms) = 2205 lb (pounds) = 1.102 sh. ton (short tons) = 0.984 long ton (long ton) API gravity degree = 141.5/relative density-131.5 (for relative density, the value at 15.5 °C is to be taken).

© The Editor(s) (if applicable) and The Author(s) 2024

L. Dou et al., *Global Oil and Gas Resources: Potential and Distribution*,

https://doi.org/10.1007/978-981-97-4756-6

References

Adams TD, Kirkby MA (1975) Estimate of world gas reserves. In: Proceedings of the 9th World Petroleum Congress, vol 3(9), pp 3–9

Ahlbrandt TS, Charpentier RR, Klett TR et al (2005) Global resource estimates from total petroleum systems. American Association of Petroleum Geologists

Bally AW (1980) Basins and subsidence: a summary. AAPG, Tulsa, Oklahoma, USA

Bian H, Tian Z, Wu Y et al (2014) Reserve growth characteristics and potential of the discovered giant oil fields in the Middle East. Petrol Explor Develop 41(2):244–247

BP (2020) BP statistical review of world energy 2020 [EB/OL]. 28 Sept 2020. https://www.bp.com/en/global/corporate/energy-economics/statistical-review-of-world-energy.html

BP (British Petroleum) (2021) BP energy outlook, 2020 edition [EB/OL]

Campbell CJ (1989) Oil price leap in the early nineties. Noroil 17(12):35–38

Campbell CJ (1997) The coming oil crisis. Multi-Science Publications Co., England, p 210

Campbell CJ (1992) The depletion of oil. Mar Pet Geol 9:666–671

CERIGAZ (2001) Natural gas in the world: 2001 survey. French Petroleum Research Institute, Paris

Committee on Geology Exploration of Petroleum of Standardization Administration of China (2011) Geological evaluating methods for tight sandstone gas: SY/T 6832. National Energy Administration, Beijing

Committee on Geology Exploration of Petroleum of Standardization Administration of China (2014) Specifications for oil sands ores exploration and estimation: SY/T 6998. National Energy Administration, Beijing

Committee on Petroleum of Standardization Administration of China (2014) Geological evaluating methods for tight sandstone gas: GB/T 30501. AQSIQ, Beijing

Committee on Petroleum of Standardization Administration of China (2015) Geological evaluating methods for shale gas: GB/T 31483. AQSIQ, Beijing

Committee on Petroleum of Standardization Administration of China (2020) Geological evaluating methods for shale oil: GB/T 38718. State Administration for Market Regulation, Standardization Administration of China, Beijing

Compilation team of PetroChina Research Institute of Petroleum Exploration and Development (RIPED) (2021) Global petroleum E&D trends and company dynamics. Petroleum Industry Press, Beijing

Compilation team of PetroChina Research Institute of Petroleum Exploration and Development (RIPED) (Chief editors: MA Xinhua, DOU Lirong, SHI Buqing, et al.) (2021) Potential and distribution of global oil and gas resources. Petroleum Industry Press, Beijing

Cross TA, Lessenger MA (1998) Sediment volume partitioning: rationale for stratigraphic model evaluation and high-resolution stratigraphic correlation. In: Sequence stratigraphy-concepts and applications. Norwegian Petroleum Society, pp 171–195

L. Dou et al., *Global Oil and Gas Resources: Potential and Distribution*,
https://doi.org/10.1007/978-981-97-4756-6

Dickinson WR (1976) Plate tectonic evolution of sedimentary basins. American Association of Petroleum Geologists Continuing Education Course Notes, Tulsa

Dickinson WR (1974) Plate tectonics and sedimentation. Special Publication, Tulsa

Dou L, Wang J, Wang R et al (2018) The Precambrian basement play in the central African rift's system. Earth Sci Front 25(2):15–23

Dou L, Wang W, Xiao W et al (2020) Progress and suggestions on CNPC's multinational oil and gas exploration and development. Petrol Sci Technol Forum 39(2):21–30

Dou L, Wei X, Wang J et al (2015) Characteristics of granitic basement rock buried-hill reservoir in Bongor Basin, Chad. Acta Petrolei Sinica 36(8):897–904+925

Dou L, Xiao K, Hu Y et al (2011) Petroleum geology and a model of hydrocarbon accumulations in the Bongor Basin, the Republic of Chad. Acta Petrolei Sinica 32(3):379–386

Dou L (2019) Exxon Mobil Corporation invests heavily in the exploration of deep water in Brazil. World Petrol Ind 26(3):71–73

EIA (2016) Review of emerging resources:US shale gas and shale oil plays [EB/OL]. 09 June 2016. https://www.eia.gov/analysis/studies/usshalegas/pdf/usshaleplays.pdf

EIA (2021) Technically recoverable shale oil and shale gas resources [EB/OL]. 07 Apr 2021. https://www.eia.gov/analysis/studies/worldshalegas/

Guo Q, Chen N, Liu C et al (2015) Research advance of hydrocarbon resource assessment method and a new assessment software system. Acta Petrolei Sinica 36(10):1305–1314

Guo Q, Xie H, Huang X et al (2016) Evaluation method system and application of oil and gas resources. Petroleum Industry Press, Beijing

Halbouty MT, Meyerhoff AA, Robert E et al (1970) World's giant oil and gas fields, geologic factors affecting their formation, and basin classification [Part 1]. In: Halbouty MT (ed) Geology of giant petroleum fields. AAPG Memoir 14, pp 502–528

Halbouty MT, Moody JD (1979) World ultimate reserves of crude oil. In: Proceedings of the 10th World Petroleum Congress, pp 291–301

He Z, Wen Z, Wang Z et al (2020) Reservoir forming plays and favorable exploration fields of Jurassic-Cretaceous in the West Siberian giant rift basin. Marine Origin Petrol Geol 25(1):70–78

Hu W, Zhang S (1992) Translation set of oil and gas resource evaluation methods (2). Information Research Institute of China National Petroleum Corporation

Hubbert MK (1969) Energy resources. In: Resources and man, pp 157–242

Hubbert MK (1956) Nuclear energy and the fossil fuels: American Petroleum Institute Drilling and Production Practice. In: Proceedings of the Spring Meeting, pp 7–25

Hubbert MK (1972) U.S. energy resources: a review as of 1972, pp 1–201

IEA (2020) World Energy Model. IEA, Paris. https://www.iea.org/reports/world-energy-model

IEA (2021) World Energy Model Documentation [EB/OL]. https://www.iea.org/reports/world-energy-model

IHS Markit (2021) IHS energy: EDIN [EB/OL]

Jarvie DM (2012) Shale resource systems for oil and gas. In: Breyer JA (ed) Shale reservoirs—giant resources for the 21st century. AAPG Memoir 97, pp 69–119

Jia C, Zou C, Li J et al (2012) Assessment criteria, main types, basic features and resource prospects of the tight oil in China. Acta Petrolei Sinica 33(3):343–350

Kaufman GM (1965) Statistical analysis of the size distribution of oil and gas fields. In: Symposium on petroleum economics and evaluation, pp 109–124

Klemme HD (1988) Petroleum basins: classifications and characteristics. J Pet Geol 3(2):1036–1058

Klett TR (2005) United States Geological Survey's reserve-growth models and their implementation. Nonrenew Resour 14(3):249–264

Klett TR, Gautier DL, Ahlbrandt TS et al (2005) An evaluation of the U.S. Geological Survey World Petroleum Assessment 2000. AAPG Bulletin, 89(8):1033–1042

Kuenen PH, Migliorini CI (1950) Turbidity currents as a cause of graded bedding. J Geol 58(2):91–127

Lei Y, Song Y, Zhang L et al (2021) Research progress and development direction of reservoir-forming system of marine gas hydrates. Acta Petrolei Sinica 42(6):801–820

Levorsen AI (1950) Estimates of undiscovered petroleum reserves. In: Proceedings of the United Nations scientific conference on the conservation and utilization of resources, pp 107–110

Levorsen AI (1956) Geology of petroleum, 1st edn. W. H. Freeman and Company, San Francisco

Li D, Shi G (2018) Optimization of common data mining algorithms for petroleum exploration and development. Acta Petrolei Sinica 39(2):240–246

Liu X, Wen Z, He Z et al (2019) Zagros basin in Middle East: along-strike variations of structure and petroleum accumulation characteristics. Acta Petrologica Sinica 35(4):1269–1278

Liu X, Zhang G, Wen Z et al (2017) Structural characteristics and petroleum exploration of Levant Basin in East Mediterranean. Petrol Explor Develop 44(4):540–548

Liu Z, Jiang X (2020) Domestic oil and gas industry development report. Petroleum Industry Press, Beijing, pp 10–11

Long S, Wang S, Sun Y et al (2005) Methods and practice of oil and gas resource evaluation. Geological Publishing House, Beijing

Margoon LB, Dow WG (1994) The Petroleum system: from source to trap. American Association of Petroleum Geologists, Tulsa

Martin AJ (1985) Prediction of strategic reserves in prospect for the world oil industry. University of Durham, pp 16–39

Masters CD, Attanasi ED, Root DH (1994) World petroleum assessment and analysis. In: Proceedings of the 14th World Petroleum Congress, pp 529–541

Masters CD, Root DH, Attanasi ED (1991) World resources of crude oil and natural gas. In: Proceedings of the 13th World Petroleum Congress, pp 51–64

Masters CD, Root DH, Turner RM (1997) World resource statistics geared for electronic access. Oil Gas J 95(41):98–104

Masters CD (1987) Global oil assessments and the search for non-OPEC oil, pp 153–169

McCabe PJ (1997) Energy resource—cornucopia or empty barrel? AAPG Bull 82(11):2110–2134

McCollough EH (1934) Structural influence on the accumulation of petroleum in California. In: Problems of petroleum geology. AAPG, Tulsa, pp 735–760

McKenzie DP (1969) Speculations on the consequences and causes of plate motions. Geophys J Roy Astron Soc 18(1):1–32

Meimer P, Slatt RM (2007) Introduction of petroleum geology of deep-water settings. AAPG Stud Geol 57:815

Nehring R (1979) The outlook for conventional petroleum resources, p 21.

Odell PR (1998) Oil and gas reserves: retrospect and prospect. Energy Explor Exploit 16:117–124

Pratt WE (1942) Oil in the Earth. University of Kansas Press, Lawrence

Rice DD (1992) Evaluation methods and application of oil and gas resources (trans. by G. Zhai et al. in Chinese). Petroleum Industry Press, Beijing

RIPED (2017) Global oil and gas E&D trends and company dynamics. Petroleum Industry Press, Beijing

RIPED (2018) Global oil and gas E&D trends and company dynamics. Petroleum Industry Press, Beijing

RIPED (2019) Global oil and gas E&D trends and company dynamics. Petroleum Industry Press, Beijing

RIPED (2020) Global oil and gas E&D trends and company dynamics. Petroleum Industry Press, Beijing

RIPED (2021) Global oil and gas E&D trends and company dynamics (2021). Petroleum Industry Press, Beijing

RIPED (2022) Global oil and gas E&D trends and company dynamics. Petroleum Industry Press, Beijing

RIPED (2023) Global oil and gas E&D trends and company dynamics. Petroleum Industry Press, Beijing

Robertson (2017) Tellus sedimentary basins of the world plays & petroleum systems: Tellus [DB/OL]. 22 Feb 2017. http://www.cgg.com/en/What-We-Do/GeoConsulting/Robertson

Schmoker JW (2002) Resource assessment perspectives for unconventional gas systems. AAPG 86(11):1993–1999

Shanmugam G (1996) High-density turbidity currents: are they sandy debris flows? J Sediment Res 66:2–10

Shell (2001) Energy needs, choices and possibilities scenarios to 2050 [EB/OL]

Styrikovich MA (1977) The long range energy perspective. Nat Res Forum 1(3):252–253

S&P Global (2021) IHS energy: EDIN [EB/OL]. 01 Jan 2019. [2021–04–31]. https://ihsmarkit.com/index.html

SPE (2016) Guidelines for application of the petroleum resources management system. [EB/OL]. [23 May 2016]. http: //www.spe.org/industry/docs/PRMS_Guidelines_Nov2011.pdf

Tao M, Liu P, Li J et al (2015) Overview and forecast of global oil and gas resources. Resour Sci 37(6):1190–1198

Tissot BP, Welte DH (1978) Petroleum formation and occurrence. Springer, Berlin

Tong X, Dou L, Tian Z et al (2003) Strategies of international petroleum exploration and development of Chinese petroleum companies in early 21st century. Petroleum Industry Press, Beijing

Tong X, He D (2001) Principles and methods of petroleum exploration. Petroleum Industry Press, Beijing

Tong X, Dou L, Tian Z (2004) Strategies of international petroleum exploration of Chinese petroleum companies. China Petrol Explor 9(1):58–64

Tong X, Li H, Xiao K et al (2009) Application of play quick analysis technique in oversea basins with low-degree exploration. Acta Petrolei Sinica 30(3):317–323

Tong X, Zhang G, Wang Z et al (2014) Distribution and potential of global oil and gas resources. Petrol Explor Develop 45(4):727–736

Tong X, Zhang G, Wang Z et al (2014) Global oil and gas potential and distribution. Earth Sci Front 21(03):1–9

USGS World Petroleum Assessment 2000 [EB/OL] (2003)

USGS (2012) An estimate of undiscovered conventional oil and gas resources of the world [EB/OL]

USGS (2005) Global resource estimates from total petroleum systems. AAPG Memoir 86

Vail PR, Mitchumjr RM, Todd RG, Widmier JM et al (1977) Seismic stratigraphy and global changes of sea-level. In: Seismic stratigraphy—applications to hydrocarbon exploration. AAPG, Tulsa, pp 49–212

Walker RG (1978) Deep-water sandstone facies and ancient submarine fans: models for exploration for stratigraphic traps. AAPG Bull 612:932–966

Wang H, Ma F et al (2016) Assessment of global unconventional oil and gas resources. Petroleum Industry Press, Beijing

Wang H, Ma F, Tong X et al (2016) Assessment of global unconventional oil and gas resources. Petrol Explor Develop 43(6):850–862

Wang Z, Wen Z, He Z et al (2021) Global condensate oil resource potential and exploration fields. Acta Petrolei Sinica 42(12):1556–1565

Wang J, Guo Q, Zhao C et al (2023) Potentials and prospects of shale oil-gas resources in major basins of China. Acta Petrolei Sinica 44(12):2033–2044

Weeks LG (1950) Discussion of "Estimates of undiscovered petroleum reserves by A.I. Levorsen". In: Proceedings of the United Nations Scientific Conference on the conservation and utilization of resources, vol 1, pp 107–110

Weeks LG (1958) Fuel reserves of the future. AAPG Bull 42(2):431–441

Weeks LG (1948) Highlights on 1947 developments in foreign petroleum fields. AAPG Bull 32:1093–1160

Weeks LG (1959) Where will energy come from in 2059? The Petroleum Engineer, pp 1A-24–A-31.

Weeks LG (1965) World offshore petroleum resources. AAPG Bull 49(10):1680–1693

Weeks LG (1971) Marine geology and petroleum resources. In: Proceedings of the 8th World Petroleum Congress, Moscow, vol 2, pp 99–106

Wen Z, Xu H, Wang Z et al (2016) Classification and hydrocarbon distribution of passive continental margin basins. Petrol Explor Develop 43(5):678–688

Wen Z, Wu Y, Bian H et al (2018) Variations in basin architecture and accumulation of giant oil and gas fields along the passive continent margins of the South Atlantic. Earth Sci Front 25(4):132–141

White D (1920) The petroleum resources of the world. Ann Am Acad Political Soc Sci 89:111–134

White IC (1885) The geology of natural gas. Science 125:521–522

Wilson TJ (1963) Hypothesis of Earth's behavior. Nature 198(4884):925–929

Wood Mackenzie. UDT (Upstream Date Tools) [D/OL]. 01 Jan 2019. [2021–04–31]. https://udt. woodmac.com/dv/

Wu S (2005) Introduction to geological evaluation of petroleum resources. Petroleum Industry Press, Beijing

Wu Y, Tian Z, Tong X et al (2014) Evaluation method for increase of reserves in large oil-gas fields based on reserves growth model & probability analysis and its application in Middle East. Acta Petrolei Sinica 35(3):469–479

Wu X, Liuzhuang X, Wang J et al (2022) Petroleum resource potential, distribution and key exploration fields in China. Earth Sci Front 29(6):146–155

Yang J, Du J, Yang Y et al (2021) Research and practice on digital transformation of the oil and gas industry. Acta Petrolei Sinica 42(2):248–258

Yu G, Xu J, Tong X, Chen M (2014) Study on the reserve growth of known oil and gas fields in the world. Earth Sci Front 21(3):195–200

Zhang D, Zhang F (2006) Three methods and their comparison of hydrocarbon resource prediction. J Northwest Univ (Nat Sci Ed) 36(3):453–456

Zhang G, Qu H, Zhao C et al (2017) Giant discoveries of oil and gas exploration in global deepwaters in 40 years and the prospect of exploration. Nat Gas Geosci 28(10):1447–1477

Zhang G, Qu H, Zhang F et al (2019) Major new discoveries of oil and gas in global deepwaters and enlightenment. Acta Petrolei Sinica 40(1):1–34

Zhao Y, Zhao Y (2019) Classification, content and extension of evaluation methods for oil and gas resources. J Southwest Petrol Univ (Sci Technol Ed) 41(4):64–74

Zheng M, Li J, Wu X et al (2019) Potential of oil and natural gas resources of main hydrocarbon bearing basins and key exploration fields in China. Earth Sci 44(3):833–847

Zou C, Tao S, Hou L et al (2011) Unconventional petroleum geology. Geological Publishing House, Beijing

Zou C, Pan S, Qhao Q (2020) On the connotation, challenge and significance of China's "energy independence" strategy. Pet Explor Dev 47(2):416–426

Zou C, He D, Jia C et al (2021) Connotation and pathway of world energy transition and its significance for carbon neutral. Acta Petrolei Sinica 42(2):233–247